AF887536

Environmental Radiochemical Analysis VI

SPONSORS

GAU Radioanalytical Laboratories

RADDEC INTERNATIONAL

GEL Laboratories UK LTD

LabLogic

ORTEC®

CAMECA®
SCIENCE & METROLOGY SOLUTIONS

nu

PYCKO SCIENTIFIC LTD
Your Alternative To The Obvious

SOCOTEC
THE POWER OF FORESIGHT

TRISKEM
SHARING INNOVATION

Environmental Radiochemical Analysis VI

Edited by

Nicholas Evans
Nottingham Trent University, UK
Email: nicholas.evans@ntu.ac.uk

ROYAL SOCIETY OF CHEMISTRY

The Proceedings of the 13th International Symposium on Nuclear and Environmental Radiochemical Analysis held at Jesus College, Cambridge, UK on 17–20 September 2018.

Special Publication No. 354

Print ISBN: 978-1-78801-735-0
PDF ISBN: 978-1-78801-773-2

A catalogue record for this book is available from the British Library

© The Royal Society of Chemistry 2020

All rights reserved

Apart from fair dealing for the purposes of research for non-commercial purposes or for private study, criticism or review, as permitted under the Copyright, Designs and Patents Act 1988 and the Copyright and Related Rights Regulations 2003, this publication may not be reproduced, stored or transmitted, in any form or by any means, without the prior permission in writing of The Royal Society of Chemistry or the copyright owner, or in the case of reproduction in accordance with the terms of licences issued by the Copyright Licensing Agency in the UK, or in accordance with the terms of the licences issued by the appropriate Reproduction Rights Organization outside the UK. Enquiries concerning reproduction outside the terms stated here should be sent to The Royal Society of Chemistry at the address printed on this page.

Whilst this material has been produced with all due care, The Royal Society of Chemistry cannot be held responsible or liable for its accuracy and completeness, nor for any consequences arising from any errors or the use of the information contained in this publication. The publication of advertisements does not constitute any endorsement by The Royal Society of Chemistry or Authors of any products advertised. The views and opinions advanced by contributors do not necessarily reflect those of The Royal Society of Chemistry which shall not be liable for any resulting loss or damage arising as a result of reliance upon this material.

The Royal Society of Chemistry is a charity, registered in England and Wales, Number 207890, and a company incorporated in England by Royal Charter (Registered No. RC000524), registered office: Burlington House, Piccadilly, London W1J 0BA, UK, Telephone: +44 (0) 20 7437 8656.

Visit our website at www.rsc.org/books

Printed in the United Kingdom by CPI Group (UK) Ltd, Croydon, CR0 4YY, UK

PREFACE

Environmental Radiochemical Analysis VI is a collection of refereed papers presented at the 13th International Symposium on Nuclear and Environmental Radiochemical Analysis, held at Jesus College, Cambridge, UK on 17–20 September 2018.

This Symposium is the latest in a long-running series of international symposia that has been held every four years by the Radiochemistry Group of the Royal Society of Chemistry, where the series now spans 50 years, and is the sixth to have its proceedings published as a book by the Royal Society of Chemistry. Its scope has changed gradually over those years and, from an initial focus on the development and application of methods for the analysis of radionuclides in environmental samples, it has now expanded to include:

- Studies of environmental source terms, transport pathways and impacts,
- Nuclear and non-nuclear industry waste characterisation and disposal
- Nuclear forensics

Eighty-one papers were presented as oral or poster presentations by delegates representing 18 different countries including the UK, with a total attendance of 130, where half were from outside of the UK.

I am grateful for the support given to this enterprise by all the members of the organising and technical committee, but special thanks goes to David Read (University of Surrey), Paul Thompson (AWE), Tony Ware, Phil Warwick (University of Southampton), Nick Evans (Nottingham Trent University), Kinson Leonard (CEFAS), and Ian Burke (University of Leeds). I am particularly grateful to Kinson and Ian for their expertise in raising sponsorship from exhibitors and controlling the finances respectively.

David Wickenden (Magnox Ltd, UK)
Chair of the Organising and Technical Committee

Contents

INDUCED RING-LIKE DEPOSITS AROUND SUSPENDED RADIOACTIVE PARTICLES BY MARANGONI STRESS AND ITS POSSIBLE USE AS DETECTION TECHNIQUE

Francisco J. Arias* and Salvador De Las Heras

Department of Fluid Mechanics, University of Catalonia, ESEIAAT C/ Colom 11, 08222 Barcelona, Spain

*francisco.javier.arias@upc.edu

ABSTRACT

In this work Marangoni-thermal convection promoted by the heat from the radioactive decay of radiogenic particles suspended in fluids is analysed. It was found that owing to the outward flow induced from the radioactive particle, the surrounding material could be pushed outwards from the particle. As a result a power-law distribution of concentration surrounding the particle appears which could reveal the presence of suspended radiogenic particles or can be used as a preliminary visual technique for detection.

1 INTRODUCTION

The object of this work was to analyse the radial convective profile surrounding a radiogenic particle suspended in a liquid medium because of the Marangoni stress induced by radial temperature profile surrounding the particle caused by its radioactive heat decay.

1.1 Statement of the Core Idea

1.1.1 Thermocapillary flow induced by temperature gradient from a suspended radiogenic particle. To begin with, let us consider a single radiogenic particle of radius r_o, which is suspended in a liquid of infinite extent, as sketched in Figure 1. The precise shape of the radiogenic particle is somewhat arbitrary. This can be highly elongated to an almost spherical shape, however, and in view of several uncertainties, a simple approximation can be made which simplifies considerably the calculations by considering two-dimensional cylindrical coordinates centred on the hot particle and assuming a cylindrical particle, let us say with radius r_o and with a length $\approx \frac{4}{3} r_o$, in such a way that the volume of the cylinder is the same than the corresponding sphere.

Because of the radiogenic heat, a thermal gradient surrounding the particle is developed with the maximum temperature at the surface of the particle and decreasing outwards. As a result, a surface tension gradient also will be developed, decreasing in the inward direction (surface tension decreases with an increase of temperature). The presence of this surface tension gradient will cause the liquid to flow away from regions of low surface tension to regions of high surface tension and then propel the surrounding suspended material away from the surface of the radiogenic particle. Bearing this conceptual framework in mind, we can proceed with a theoretical treatment of the phenomenon.

Figure 1 *Physical model of cylindrical particle suspended into a liquid.*

First, owing to the symmetry of the problem, only the radial component of the velocity v_r and the longitudinal velocity v_z are not null. Furthermore, far away from the particle $v_z \ll v_r$ and so it can also be neglected. With these simplifications, the radial component of the Navier-Stokes equation and the equation of continuity for steady motion are given by,[1]

$$v_r \frac{\partial v_r}{\partial r} = -\frac{1}{\rho}\frac{\partial p}{\partial r} + \nu\left(\frac{\partial^2 v_r}{\partial r^2} + \frac{\partial^2 v_r}{\partial z^2} + \frac{1}{r}\frac{\partial v_r}{\partial r} - \frac{v_r}{r^2}\right) \qquad (1)$$

$$\frac{\partial v_r}{\partial r} + \frac{v_r}{r} = 0 \qquad (2)$$

where v_r is the radial velocity of the liquid, p is the pressure, and ρ and ν the liquid density and kinematic viscosity, respectively.

By combining Eq.(1) and Eq.(2) one obtains:

$$v_r \frac{\partial v_r}{\partial r} = -\frac{1}{\rho}\frac{\partial p}{\partial r} + \nu\frac{\partial^2 v_r}{\partial z^2} \qquad (3)$$

Because thermocapillary flow is always slow enough, the convective term $v_r \frac{\partial v_r}{\partial r}$ which is proportional to $\approx v_r^2$, can be considered as being of second order in comparison with the other terms, and thus can be neglected (linear approach), Thus, Eq.(3) simplifies as:

$$\frac{\partial p}{\partial r} = \mu \frac{\partial^2 v_r}{\partial z^2} \quad (4)$$

where μ is the dynamic viscosity of the liquid. Because the surface motion of the viscous liquid return backwards due to the container walls, this causes the appearance of a pressure drop along the radial axis. Therefore, the pressure of the liquid at any point surrounding the particle can be expressed formed by two compounds,[2]

$$p = \rho g z + p_r(r) \quad (5)$$

Likewise, conservation of mass requires that:

$$v_r(r, z) = \frac{F(z)}{r} \quad (6)$$

where $F(z)$ is a function which depends only of the depth z. Thus, Eq.(4) becomes:

$$\frac{dp(r)}{dr} = \frac{\mu}{r} \frac{d^2 F(z)}{dz^2} \quad (7)$$

Since $p(r)$ does not depends on the coordinate z, and likewise, $F(z)$ does not depend on the coordinate r, therefore, solving Eq.(4) we get:

$$v_r(r, z) = \frac{1}{r}\left[C_1 + C_2 z + \frac{z^2}{2\mu} \frac{dp_r(r)}{dr}\right] \quad (8)$$

where C_1 and C_2 are constants to be determined by imposed boundary conditions. The boundary conditions for our system are given by:

$$v_r(r, z = h) = 0 \quad (9)$$

where zero slip on the solid boundary is assumed. On the other hand, the condition at the free liquid interface where the viscous-stress tensor must be continuous, and then the continuity of p_rz at the free-surface where z=0, implies,[2]

$$\frac{\partial \sigma}{\partial r} = -\mu \frac{\partial v_r(r,z)}{\partial z} \quad (10)$$

Taking into account the boundary conditions given in Eq.(9) and Eq.(10) into Eq.(8) we obtain:

$$C_1 = -\frac{r}{\mu} \frac{d\sigma}{dr} \quad (11)$$

and

$$C_2 = \frac{rh}{\mu}\frac{d\sigma}{dr} - \frac{h^2}{2\mu}\frac{dp_r(r)}{dr} \tag{12}$$

which inserting into Eq.(8) yields:

$$v_r(r,z) = \frac{1}{r}\left[\frac{r}{\mu}\frac{d\sigma}{dr}(h-z) - \frac{1}{2\mu}\frac{dp_r(r)}{dr}(h^2 - z^2)\right] \tag{13}$$

The surface tension gradient induced by the thermal gradient surrounding the radiogenic particle may be expressed as:

$$\frac{d\sigma}{dr} = \frac{d\sigma}{dT}\frac{dT}{dr} \tag{14}$$

and on the other hand the flow of heat by conduction is governed by the Fourier equation:

$$-\kappa_e\frac{dT}{dr} = q(r)A_r \tag{15}$$

where $q(r)$ is the heat flux at which heat is conducted in the radial direction through a plane of area $A_r = 2\pi r dz$ normal to this direction at a distance r where the temperature gradient is $\frac{dT}{dr}$; and κ_e is the effective thermal conductivity of the liquid which includes both conduction and convection and is defined as,[3]

$$\kappa_e = \kappa Nu \tag{16}$$

where κ is the thermal conductivity of liquid, and Nu is the Nusselt number. For our case, the heat transfer from the reduced thermocapillary flow is much lower than the conductive heat transfer and then $Nu \simeq 1$ with $\kappa_e \simeq \kappa$. On the other hand, from an energy balance we know that the heat flux at a distance r is given by $q_o 2\pi r_o = q(r)2\pi r$ where r_o is the radius of the particles and q_o the heat flux at its surface, respectively. Thus, Eq.(15) becomes:

$$-\kappa_e\frac{dT}{dr} = \frac{q_o r_o}{r} \tag{17}$$

Substituting Eq.(17) into Eq.(14) one obtains:

$$\frac{d\sigma}{dr} = -\frac{d\sigma}{dT}\frac{q_o}{\kappa_e}\frac{r_o}{r} \tag{18}$$

and inserting Eq.(18) into Eq.(13) yields:

$$v_r(r,z) = \frac{1}{r}\left[-\frac{q_o r_o}{\mu\kappa_e}\frac{d\sigma}{dT}(h-z) - \frac{1}{2\mu}\frac{dp_r(r)}{dr}(h^2 - z^2)\right] \tag{19}$$

The only parameter remaining to be evaluated is $\frac{dp_r(r)}{dr}$. This may be done by considering that the average mass flow over a cylindrical cross section must be equal to zero.

$$\int_0^h v_r(r,z)2\pi r dz = 0 \tag{20}$$

For convenience, it is easier to split the integral into two regions as:

$$\int_0^{h_1} v_r(r,z) 2\pi r dz + \int_{h_1}^{h} v_r(r,z) 2\pi r dz = 0 \qquad (21)$$

where h_1 is the region from the surface to a depth approximated to the size of the particle (i.e., $h_1 \approx r_o$). This split of the integral is useful because the thermocapillary flow is only appreciable to very small depths from the surface, and then the second term in Eq.(21) is practically nil. Thus, Eq.(21) becomes:

$$\int_0^{h_1} v_r(r,z) 2\pi r dz \approx 0 \qquad (22)$$

by inserting Eq.(19) into Eq.(22) and after integrating we obtain:

$$v_r(r,z) = -\frac{q_o r_o}{4\mu\kappa_e h_1} \frac{d\sigma}{dT} (3z^2 - 4h_1 z + h_1^2) \cdot \frac{1}{r} \qquad (23)$$

Figure 2 shows a sketch of the predicted velocity profile from the induced thermocapillary motion. Finally, the surface velocity is given when $z = 0$ and yields:

$$v_r(r) = -\frac{q_o r_o h_1}{4\mu\kappa_e r} \frac{d\sigma}{dT} \qquad (24)$$

If it is desired, the heat flux at the surface of a cylindrical particle may be related to its radiogenic activity by the following expression:

$$q_o \simeq \frac{\varepsilon\rho_p \eta N_A \lambda r_o}{2\overline{m}_p} \qquad (25)$$

where ε is the energy per disintegration, ρ the density of particle, η its mineral content (fraction of radiogenic material), N_A the Avogadro number, λ the disintegration constant, $\overline{m}_p$ the molecular weight and r_o the radius of the particle. Substituting Eq.(25) into Eq.(24) one obtains:

$$v_r(r) = -\frac{\varepsilon\rho_p \eta N_A \lambda r_o^2 h_1}{8\overline{m}_p \mu\kappa_e r} \frac{d\sigma}{dT} \qquad (26)$$

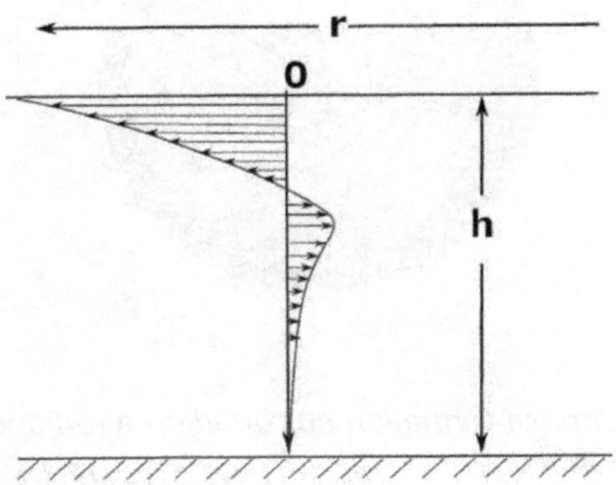

Figure 2 *Velocity profile for the thermocapillary motion over a cylindrical cross section.*

1.2 Ring Formation

From the calculated velocity profile we can derive an expression for the formation of the diffusive ring as follows.

First, let us assume that the suspended radiogenic particle is surrounded by a cloud of particles homogeneously dispersed in the liquid. Let us also assume that the container has a radius R, for example the radius of the test tube used in the laboratory. Because of thermocapillary flow, particles are pushed away from the radiogenic particle and as result a radial distribution of particles will appear as depicted in Figure 3.

The flux of particles J across any plane parallel to the z-axis is given by:

$$J = cv_r \tag{27}$$

where c is the total concentration of particles at a given radial distance r, and v_r is the radial velocity which was previously calculated. On the other hand, the opposite flux due to particle concentration gradient is given by:

$$J_p = -D_p \frac{dc}{dr} \tag{28}$$

where D_p is the local particle diffusivity.

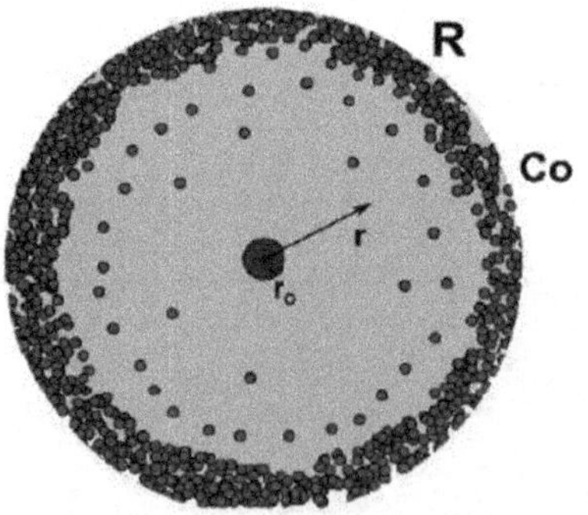

Figure 3 *Ring-like deposit formation surrounding a radiogenic particle induced by thermocapillary motion.*

At steady state, this flux must be equated to the flux across any plane parallel to the z-axis, and then with $J_p = -J$, from Eq.(27) and Eq.(28) we obtain:

$$\frac{dc}{c} = \frac{v_r}{D_p} dr \tag{29}$$

inserting Eq.(26) into Eq.(29) and if we define the concentration of particles at the wall container as, say, c_o, after integration one obtains the following power-law for the concentration:

$$\frac{c}{c_o} = \left[\frac{r}{R}\right]^{-n} \tag{30}$$

here an exponential index n is defined as:

$$n = \frac{\varepsilon \rho_p \eta N_A \lambda r_o^2 h_1}{8 D_p \overline{m}_p \mu \kappa_e} \frac{d\sigma}{dT} \tag{31}$$

Due to the very low velocity regime, it is permissible to use the Einstein-Stokes diffusivity equation for the diffusivity term which is valid for low Reynolds numbers and is given by:

$$D_p = \frac{\kappa_B T}{6\pi\mu a} \tag{32}$$

where κ_B is the Boltzmann constant, T the temperature, and a the radius of the particle suspended surrounding the radiogenic particle. Inserting Eq.(32) into Eq.(31) yields:

$$n = \frac{3\pi a \varepsilon \rho_p \eta N_A \lambda r_o^2 h_1}{4 \kappa_B T \overline{m}_p \kappa_e} \frac{d\sigma}{dT} \tag{33}$$

In addition, a mild simplification may be assumed considering that the depth h_1 in which the flow is becoming almost null, is of the same order than the length scale of the particle, i.e., $h_1 \approx r_o$. Therefore the exponential index given in Eq.(33) yields:

$$n \approx \frac{3\pi a \varepsilon \rho_p \eta N_A \lambda r_o^3}{4 \kappa_B T \overline{m}_p \kappa_e} \frac{d\sigma}{dT} \tag{34}$$

2 SUMMARY OF RESULTS AND CONCLUSIONS

According with our previous discussion, the profile concentration surrounding a radiogenic particle is given by a power-law, Eq.(30), with an exponential index defined by Eq.(34). The exponential index is directly related to the decay constant λ and the radioactive content of the particle η, and therefore, the formation of such rings can be a kind of "fingerprint" betraying the presence of radiogenic particles suspended in the liquid. Another application could be in geology. In fact, the presence of "frozen rings" in rocks could be an indication that in the past - when the rock was molten - radiogenic particles were present in the rock or even that the driving heat was radiogenic.

3 NOMENCLATURE

A_r = Area normal to the radial direction

C_1 = constant

C_2 = constant

D_p = diffusivity

$F(z)$ = component radial velocity dependent of depth

g = gravity

h = depth of the container

h_c = heat transfer coefficient.

l = characteristic length

$\bar{m}_p$ = molecular weight of radiogenic particle

n = dimensionless exponent, Eq.(34)

N_A = Avogadro number

Nu = Nusselt number

p = pressure

q_o = heat flux at the surface of particle

q = heat flux at a distance r

r = radial coordinated directed from the particle outwards

r_o = radius of radiogenic particle

R = radius of the container

T = temperature

v = velocity

z =the z-axis coordinated directed
from the surface to the solid boundary

3.1 Greek symbols

η = radiogenic mineral content of particle

ε = energy per disintegration

λ = disintegration constant

ν = kinematic viscosity

μ = dynamic viscosity

ρ = density of liquid

ρ_p = density of radiogenic particle

σ = surface tension

$\boldsymbol{\kappa}$ = thermal conductivity of liquid

$\boldsymbol{\kappa_e}$ = effective thermal conductivity

3.2 Subscripts symbols

$\boldsymbol{e}$ = effective

$\boldsymbol{p}$ = particle

$\boldsymbol{r}$ = radial

z = vertical coordinate

CONFLICTS OF INTEREST

There are no conflicts to declare.

ACKNOWLEDGEMENTS

This research was supported by the Spanish Ministry of Economy and Competitiveness under fellowship grant Ramon y Cajal: RYC-2013-13459.

REFERENCES

1. L. D. Landau, E. M. Lifshiftz, *Fluid Mechanics,* Pergamon, first published in English, 1959.
2. A. I. Fedosov, Thermocapillary Motion, *Zhurnal Fizicheskoi Khimii.*, 1956, **30**(2), 366–373.
3. Y. A. Cengel, *Heat and Mass Transfer: Fundamentals & Applications*, 5th edn.

PRODUCTION AND CHARACTERISATION OF REFERENCE MATERIALS IN SUPPORT OF NATURALLY OCCURRING RADIOACTIVE MATERIAL (NORM) INDUSTRIES

Emma Braysher,*[a,b] Ben Russell,[b] Franck Dal Molin [c] and David Read [a,b]

[a] Chemistry Department, University of Surrey, Stag Hill, Guildford, Surrey, GU2 7XH; [b] Nuclear Metrology Group, National Physical Laboratory, Hampton Road, Teddington, Middlesex, TW11 0LW; [c] TATA Steel, Swinden Technology Centre, Rotherham, Yorkshire, S60 3AR.

*emma.braysher@npl.co.uk

ABSTRACT

Naturally occurring radioactive material is a problematic by-product of a range of industries and needs to be handled, stored, processed and disposed of in a safe and economic manner. Accurate characterisation of such material should be underpinned by measurement of certified reference materials in order to validate the methods employed and ensure quality control. This work highlights the current shortage of suitable reference materials and the approach being followed to address this issue, initially for the steel and oil and gas industries.

1 INTRODUCTION

1.1 Naturally occurring radioactive materials

Naturally occurring radioactive materials (NORM) are comprised of diverse matrices containing natural series isotopes at concentrations that have been enhanced by anthropogenic processes[1]. Thorium-232 ($t_{½}$ = 1.40 × 10^{10} a), ^{238}U ($t_{½}$ = 4.468 × 10^{9} a) and ^{235}U ($t_{½}$ = 7.04 × 10^{8} a) are the major long-lived radioisotopes with corresponding decay chains that are significant in NORM. Isotopic fractionation accompanies chemical separation processes, typically leading to enhanced levels of the radium isotopes along with ^{210}Pb and ^{210}Po[2].

The NORM-producing industries, as defined in the Environmental Permitting Regulations 2018 (UK)[3] and the Euratom Basic Safety Standards of the

European Commission[4] are shown in Table 1. The matrix of a NORM residue and the radionuclides present will vary depending on the origin of the raw materials and their processing history. Table 2 illustrates the main NORM industries relevant to the UK, detailing their primary matrices and the most important radionuclides present. The limit for NORM waste to be disposed of in un-assessed landfill in the UK is 5 Bq/g[5]; thus, accurate characterisation methods are required to ensure safe and cost-effective handling, recycling, storage or disposal of both raw materials and by-products.

Table 1 *The NORM-producing industries as identified by UK (Environmental Permitting Regulations 2018) and EU (Euratom Basic Safety Standards of the European Commission) regulations.*[3,4]

NORM industry	Regulations	
	EU	UK
Production and use of thorium	☐	☐*
Oil and gas production	☐	☐
Thermal phosphorus production	☐	☐
Zircon and zirconium industry	☐	☐
Extraction of rare earth elements	☐	☐
Mining and processing of ores other than uranium	☐**	☐
Titanium dioxide pigment production	☐	☐
Smelting of tin, copper, aluminium, zinc, lead, iron and steel	☐	☐
Processing of niobium/tantalum ore	☐	
Geothermal energy production	☐	
Cement production	☐	
Coal fired power plants	☐	
Production of phosphoric acid and phosphate fertilisers	☐	
Primary iron production	☐	
Ground water filtration facilities	☐	
Production and use of uranium		☐*
Removal and management of radioactive scales		☐
Coal mine de-watering plants		☐
China clay extraction		☐
Water treatment for drinking water		☐

*In the UK uranium and thorium production are described as "type 1 NORM industrial activities". All others are described as "type 2 NORM industrial activities".

**Specifically niobium and tantalum ore.

Table 2 *Main NORM industries relevant to the UK and their corresponding NORM matrices.*

NORM industry	Matrix	Composition	Radionuclides of interest	Use	Ref
Oil and gas production	Scale	Barite ($BaSO_4$), metallic Pb, litharge (PbO), cerussite ($PbCO_3$), galena (PbS)	$^{226/228}Ra$, ^{210}Pb, ^{210}Po	Waste	7
	Sludge	Barite	$^{226/228}Ra$	Waste	14
Titanium dioxide pigment production	Scale	Rutile and anatase (TiO_2), barite	$^{226/228}Ra$	Waste	7
	Sludge	Barite	^{232}Th, $^{226/228}Ra$	Waste	15
Cement production	Blast furnace slag	Iron oxides, calcium, carbon	^{226}Ra, ^{232}Th, ^{40}K	Cement feedstock from steel industry	16
	Coal fly ash	SiO_2, Al_2O_3, Fe_2O_3, CaO	$^{226/228}Ra$, ^{232}Th, ^{40}K	Cement feedstock from coal-fired power plants	16
	Red mud	Fe_2O_3, Al_2O_3, TiO_2, CaO, SiO_2, Na_2O	$^{226/228}Ra$, ^{232}Th, ^{40}K	Cement feedstock from aluminium production	16
Production of phosphoric acid and phosphate fertilisers	Phosphogypsum	Gypsum ($CaSO_4 \cdot 2H_2O$)	$^{226/228}Ra$	By-product used as feedstock in construction industry	13
Steel industry	Blast furnace slurry	Iron oxides, calcium, carbon	^{210}Po, ^{210}Pb	Waste	17
	Blast furnace slag	Iron oxides, calcium, carbon	$^{226/228}Ra$, ^{232}Th, ^{40}K	By-product used as feedstock for cement manufacture	18
	Precipitated dusts	Iron oxides, calcium, carbon	^{210}Po, ^{210}Pb	Recycled	17
China clay extraction	Scale	Barite	$^{226/228}Ra$, ^{210}Pb, ^{210}Po	Waste	19

1.2 Certified reference materials

Certified reference materials (CRM) are used to ensure quality control and metrological traceability through method validation and calibration of instruments[6]. Similarity in composition between a CRM and the material being analysed is vital for the accuracy and precision of the measurement. Owing to the wide range of sample types across industrial sectors, the lack of CRM for NORM is a huge constraint[6–12]. Currently, the only available NORM reference material is for a phosphogypsum[13].

Dal Molin *et al*[7] proposed establishing a generic set of CRM for the primary NORM matrices (barite, rutile, silicate, iron oxide and lead-based), allowing validation of the analytical process, supplemented by bespoke radioactive tracers appropriate to each set of measurements. The disadvantage of such an approach, as acknowledged by the authors, is that the radioactive 'spikes' may not be representative of nuclides in refractory phases.

This work aims to develop bespoke CRM for selected NORM materials of major importance to the UK, initially focussing on steel production and the oil & gas industry. The materials will be certified for the activity of a selection of radionuclides through an inter-laboratory comparison exercise, which is being formulated for this work and to assist testing laboratories in the UK and elsewhere. The inter-comparison exercise will take into account measurements from approximately 24 participating laboratories, using their routine methods to find the activity of selected radionuclides reported with 2σ uncertainty. The elemental composition of the materials will be investigated using inductively coupled plasma mass spectrometry (ICP-MS) and x-ray fluorescence spectrometry (XRF), following the exercise.

2 MATERIALS

2.1 Blast furnace slag

Blast furnace slag (BFS) is one of the primary by-products from the steel production process. It is an amorphous material with high silicon and calcium content, which is commonly recycled and used as a replacement for Portland cement in the building industry. Owing to its use as a building material, there is a resulting risk of indoor radiation exposure[16,20,21] and, consequently, the European Commission has proposed radiation hazard indices for gamma and alpha exposures (Equation 1 & 2).

$$I_\gamma = \left(\frac{A_{Ra}}{300} + \frac{A_{Th}}{200} + \frac{A_K}{3000}\right) \quad (1)$$

$$I_\alpha = \left(\frac{A_{Ra}}{200}\right) \quad (2)$$

Materials of $I_\gamma > 1$ should not be used in construction as occupants of such buildings will receive a dose rate higher than 1 mSv y^{-1}, which is the public exposure limit set by the European Comission[4,22]. The alpha index gives I_{α} = *1* for a ^{226}Ra concentration of 200 Bq kg^{-1}, which is the recommended maximum in building materials[22]. With these indexes taken into consideration, the radionuclides of interest in the blast furnace slag are ^{226}Ra, ^{232}Th and ^{40}K.

2.2 Oil pipe scale

The oil and gas industry includes exploration, extraction and refining of oil and gas to produce petroleum products. The rock formations in which the oil is found contain uranium and thorium, leading to the presence of radionuclides in any excavated material, including scale build-up in pipelines. The scale can block pipes and therefore must be removed, creating NORM waste[19,23,24]. The scale in this study is a mineral scale rich in barium sulphate (barite, $BaSO_4$), which co-precipitates radium. The radionuclides of interest for scale material are therefore ^{226}Ra and ^{228}Ra, as well as ^{210}Po and ^{210}Pb. This is contrary to a metallic scale, in which ^{210}Po and ^{210}Pb levels would be enhanced.

3 METHODS AND RESULTS

3.1 Preparation of reference material

The process of creating a reference material must follow certain guidance to make sure it is traceable and its certification is valid. In this study, ISO guide 35:2017 was followed, providing guidance for characterisation and assessment of homogeneity and stability[25]. An outline of the process is shown in Fig. 1. The materials are ground and sieved to <150 µm using sieves of different pore sizes and a sieve shaker. Materials are dried in an oven at 150°C and then mixed in a drum mixer overnight.

Equations 3-9 can be used to express the uncertainty of the consensus activity value of the reference material, taking into account the homogeneity and stability of the samples;

$$u_{bb} = \sqrt{\frac{\sum_i (x_i - \overline{x_i})^2}{n-1}} \left(\frac{1}{\overline{x_i}}\right) \tag{3}$$

$$u_{meas} = \sqrt{\frac{\sum_j (x_j - \overline{x_j})^2}{m-1}} \left(\frac{1}{\overline{x_j}}\right) \tag{4}$$

$$u_{int} = mean\left(\frac{u_i}{x_i}\right) \tag{5}$$

$$u_{hom}^2 = u_{bb}^2 - u_{meas}^2 \tag{6}$$

or (whichever gives the lower result);

$$u_{hom}^2 = u_{bb}^2 - u_{int}^2 \tag{7}$$

$$u_{N,rel}^2 = u_{cons}^2 + u_{hom}^2 + u_{stab}^2 \tag{8}$$

$$u_N = u_{N,rel} N \tag{9}$$

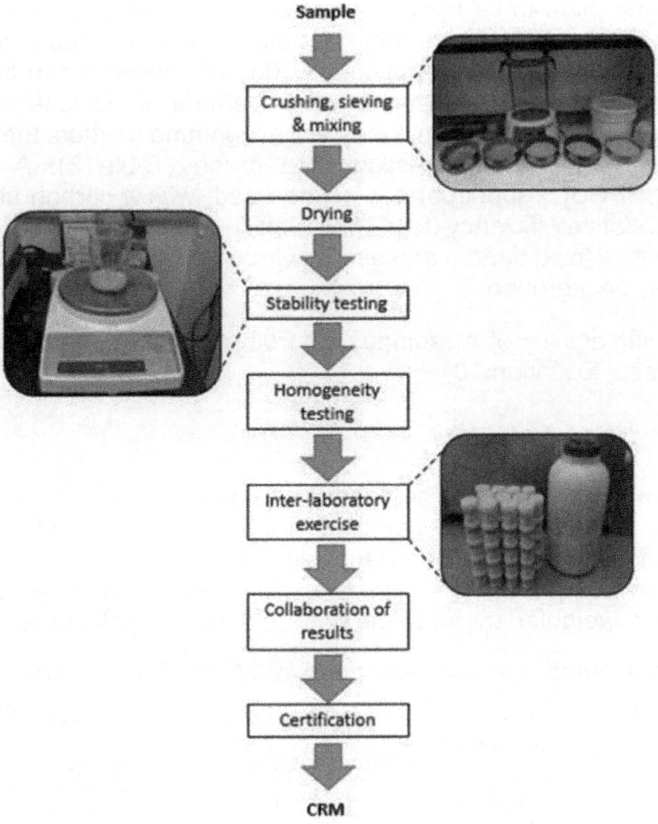

Figure 1 *The process of preparation of a reference material based on ISO guide 35:2017. Images show progress for certification of blast furnace slag material.*

where n is the number of samples measured, x_i and x_j are the decay-corrected count rate per unit mass of sample i and j respectively (cps g^{-1}), u_{bb} is the relative standard deviation of x_i, m is the number of measurements on a single selected sample, u_i is the standard uncertainty of x_i (cps g^{-1}), u_{meas} is the relative measurement uncertainty, u_{int} is the mean of the relative uncertainties of x_i, u_{cons} is the standard uncertainty of the consensus value, u_{hom} is the relative homogeneity uncertainty, u_{stab} is the relative stability uncertainty, N is the assigned specific activity value, $u_{N,rel}$ *is* the relative uncertainty of the assigned value N, and u_N is the uncertainty of the assigned value N[10].

3.2 Gamma spectrometry

A number of the radionuclides of interest to NORM industries undergo gamma decay, allowing them to be identified and quantified by gamma spectrometry. This technique allows measurement of multiple radionuclides simultaneously with limited or no sample preparation. NORM radionuclides can be measured by their gamma emissions, or by the gamma emissions of daughter radionuclides in the decay chain if they are not gamma emitters themselves, as long as the appropriate corrections are made (Table 3). A high purity germanium (HPGe) γ-spectrometer will be used, with a carbon fibre detector window at a relative efficiency (to a 3″×3″ NaI(Tl) detector) of 65 %. A cylindrical lead shield with a fixed bottom and a movable cover will shield the detector from external γ-ray background.

The specific activity of a radionuclide from a certain gamma peak can be calculated using Equation 10;

$$A_{Ei} = \frac{N_{Ei}}{\varepsilon_E \times t \times \gamma_d \times M_s} \qquad (10)$$

where NEi and εE are the net peak area and detection efficiency of a peak at energy E, t is the counting live time, γ_d is the number of gammas per disintegration of this nuclide for a transition at energy E, and Ms is the mass of the measured sample in kg. If there is more than one peak for a certain radionuclide, a weighted mean of the nuclide activity is calculated[26].

Table 3 *Characteristic gamma emissions for NORM radionuclides.*

Radionuclide	Gamma emission	Energy (keV)	Probability (%)
^{238}U	^{214}Pb	241.98	7.268
	^{214}Pb	295.21	18.41
	^{214}Pb	351.83	35.60
	^{214}Bi	609.31	45.49
	^{214}Bi	768.36	4.892
	^{214}Pb	785.91	1.06
	^{214}Bi	806.17	1.262
	^{214}Bi	934.06	3.10
^{232}Th	^{212}Pb	238.6	43.6
	^{228}Ac	338.3	11.4
	^{228}Ac	911.2	26.2
	^{228}Ac	968.9	15.9
^{226}Ra	^{226}Ra	186.18	3.555
^{40}K	^{40}K	1460.82	10.55

Losses from attenuation by the surrounding material and the sample itself must be taken into account during measurement, as must the efficiency of the detector and the geometry used. Attenuation will affect gamma emissions of lower energy more readily than that of higher energy, and therefore an efficiency calibration, in addition to a standard energy calibration, must be used. The detection efficiency for a peak is determined using Equation 11;

$$\varepsilon_E = \frac{N_E}{A \times t \times P_\gamma} \qquad (11)$$

where the components are as above, A is the source activity in Bq, and $P\gamma$ is the absolute probability of an emission.

4 CONCLUSIONS

To help to address the critical lack of CRM in NORM industries, two reference materials are under development, following the procedures outlined in ISO guide 35:2017. Subsequent stability and homogeneity testing is being undertaken to contribute ustab and uhom to the overall uncertainty. This will be followed by a global inter-laboratory exercise to find a consensus value for the specific activity, N. The overall uncertainty (uN) will be determined and will take into account the standard uncertainty of the consensus value ucons. These materials will aid characterisation through validation of methods and calibration of instruments, ensuring safety and environmental protection in its handling, reprocessing and disposal.

CONFLICTS OF INTEREST

The authors have no conflicts of interest to declare.

ACKNOWLEDGEMENTS

This work is funded jointly by the University of Surrey and the National Physical Laboratory. The National Physical Laboratory is operated by NPL Management Ltd, a wholly-owned company of the Department for Business, Energy and Industrial Strategy (BEIS).

REFERENCES

1 M. S. Al-Masri and A. Aba, *Appl. Radiat. Isot.*, 2005, **63**, 457–463.

2 D. Desideri, M. A. Meli, L. Feduzi and C. Roselli, *Int. J. Environ. Anal. Chem.*, 2006, **86**, 601–613.

3 DEFRA, *UK Statutory Instrument No. 0000: The Environmental Permitting (England and Wales (Amendment) Regulations 2018*, National Radiological Protection Board, 2018.

4 European Commission, *Off. J. Eur. Union*, 2013, **L13/1**, 242.

5 DEFRA, *Exemption Guidance for NORM industrial activities on how to comply with the radioactive substances exemption regime: The Environmental Permitting Regulations 2011 (Amendment)*, 2013.

6 S. Parry, *Radiochim. Acta*, 2012, **100**, 495–501.

7 F. Dal Molin, P. E. Warwick and D. Read, *Anal. Chim. Acta*, 2018, **1000**, 67–74.

8 C. Larijani, A. K. Pearce, P. H. Regan, B. C. Russell, S. M. Jerome, M. T. Crespo, P. de Felice, G. Lutter, F. Maringer and M. Mazánová, *Appl. Radiat. Isot.*, 2017, **126**, 279–284.

9 F. Josef Maringer, A. Baumgartner, F. Cardellini, P. Cassette, T. Crespo, J. Dean, H. Wiedner, J. Hůlka, F. Kabrt, P. Kovář, C. Larijani, G. Lutter, M. Marouli, A. Mauring, M. Mazánová, B. Michalik, N. Michielsen, V. Peyres, S. Pierre, R. Pöllänen, S. Pommé, M. Reis, M. Stietka, L. Szücs and B. Vodenik, *Appl. Radiat. Isot.*, 2017, **126**, 273–278.

10 A. Harms and C. Gilligan, *Appl. Radiat. Isot.*, 2010, **68**, 1471–1476.

11 S. M. Judge and P. H. Regan, *Radiat. Phys. Chem.*, 2017, **140**, 463–465.

12 S. Carter, A. Fisher, B. Gibson, J. Marshall, B. Russell and I. Whiteside, *J. Anal. At. Spectrom.*, 2017, **32**, 2068–2117.

13 A. Shakhashiro, U. Sansone, H. Wershofen, A. Bollhöfer, C. K. Kim, C. S. Kim, G. Kis-Benedek, M. Korun, M. Moune, S. H. Lee, S. Tarjan and M. S. Al-Masri, *Appl. Radiat. Isot.*, 2011, **69**, 231–236.

14 J. Garner, J. Cairns and D. Read, *J. Environ. Radioact.*, 2015, **150**, 49–56.

15 J. Mantero, M. J. Gazquez, J. P. Bolivar, R. Garcia-Tenorio and F. Vaca, *J. Environ. Radioact.*, 2013, **120**, 26–32.

16 I. García-Díaz, M. J. Gázquez, J. P. Bolivar and F. A. López, *Characterization and Valorization of Norm Wastes for Construction Materials*, InTech, 2016.

17 F. Dal-Molin, E. Aries, R. Fisher, D. Anderson and S. Johnston, in *CETAS 8th International Workshop on progress in analytical chemistry & materials characterisation in the steel and metal industries*, Luxembourg, 2011, pp. 315–323.

18 S. Alberici, J. De Beer, I. Van Der Hoorn and M. Staats, *Fly ash and blast furnace slag for cement manufacturing: BEIS research paper no. 19*, 2017.

19 D. Read, B. Rabey, S. Black, F. . Glasser, C. Grigg and A. Street, *Miner. Eng.*, 2004, **17**, 293–304.

20 R. Trevisi, F. Leonardi, S. Risica and C. Nuccetelli, *J. Environ. Radioact.*, 2018, **187**, 90–105.

21 N. Todorovic, J. Nikolov, M. Krmar, D. Mrda, S. Todorovic and P. Kuzmanovic, in *Proceedings of the Tenth Symposium of the Croatian Radiation Protection Association : HDZZ - CRPA : Zagreb 2015*, Hrvatsko društvo za zaščitu od zračenja, Zagreb, 2017.

22 E. Bavarnegin, M. V. Moghaddam and N. Fathabadi, *J. Med. Phys.*, 2013, **38**, 93–7.

23 B. Heaton and J. Lambley, *Appl. Radiat. Isot.*, 1995, **46**, 577–581.

24 M. S. Al-Masri and H. Suman, *J. Radioanal. Nucl. Chem.*, 2003, **256**, 159–162.

25 BSI, *PD ISO Guide 35:2017 Reference materials - guidance for*

characterization and assessment of homogeneity and stability, 2017.
26 M. Tzortzis, H. Tsertos, S. Christofides and G. Christodoulides, *Radiat. Meas.*, 2003, **37**, 221–229.

INTER LABORATORY COMPARISON EXERCISE TO DETERMINE THE RADIONUCLIDE COMPARISON OF THREE MIXED BIOTA SAMPLES

Jane Caborn[a*], Paul Blowers[b], Roger Benzing[c], Alison Boler[d], John Cobb[e], Alastair Dewar[b], Christopher Hardy[f], David Harris[g], Anthony Lees[h], Lorna Mitchell[i], Máirín O'Colmáin[j], David Tait[k], Andrey Volynkin[l] and Phillip. E. Warwick[m]

[a] Tractebel, Level 20, 25 Canada Square, London, E14 5LQ, UK, [b] Cefas Lowestoft Laboratory, Pakefield Road, Lowestoft, Suffolk,NR33 0HT,UK, [c] SCOTEC, 12 Moorbrook, Didcot OX11 7HP, UK, [d] DSTL - Porton Down, Idmiston, Salisbury SP4 0JG, UK, [e] Wood Nuclear Limited, Renaissance Centre, 601 Faraday Street, Birchwood Park, Warrington, WA3 6GN, UK, [f] Marine & Technology, Babcock International Group, Rosyth Business Park, Rosyth, Dunfermline Fife, KY11 2YD, UK, [g] Atkins, Hunterston B power station, West Kilbride, Ayrshire, KA23 9QX, UK, [h] Cavendish Nuclear Limited, Greeson Court, Westlakes Science & Technology Park, Crow Park Way, Moor Row, Cumbria, CA24 3HZ, UK, [i] CRCE, Radiation Hazards and Emergencies Dept., Public Health England, 155 Hardgate Road, Glasgow, G51 4LS, UK, [j] EPA Regional Inspectorate Dublin, McCumiskey House, Richview, Clonskeagh Road, Dublin 14, Ireland, [k] Max Rubner-Institut (MRI), Bundesforschungsinstitut für Ernährung und Lebensmittel, Institut für Sicherheit und Qualität bei Milch und Fisch, Hermann-Weigmann-Straße 1, D-24103 Kiel, Germany, [l] Havforskningsinstituttet/Institute of Marine Research, P. B. 1870 Nordnes, N-5817 Bergen, Norway, [m] GAU-Radioanalytical, National Oceanography Centre, European Way, Southampton, SO14 3ZH, UK

*cabornjane@gmail.com

ABSTRACT

The issue surrounding the shortage of reference materials has long been discussed and becoming progressively important as the nuclear industry faces the challenges of decommissioning, increasing the requirement for the analysis of a wide range of radionuclides in diverse sample types. To meet demand the radiochemical methods need to be flexible and robust whilst maintaining a high level of quality, essential to underpinning public and environmental safety in a highly regulated sector. Integral to the process of method development, validation, and performance evaluation is the use of reference materials which provide the confidence in the measurement. Currently there is an acute shortage of certified, solid, reference materials due to costs associated with production and availability of suitable material.An alternative to adopting the formal intercomparison exercise approach was explored by the UK's Analyst Informal Working Group (AIWG). The members recognized the shortage of suitable, solid materials containing a range of alpha, beta and gamma emitters and identified three mixed biota samples. The molluscs, winkles and crustacea contained a mix of environmental activity level radionuclides (^{14}C, ^{40}K, ^{90}Sr, ^{99}Tc, ^{137}Cs, $^{239+240}Pu$, ^{241}Pu and ^{241}Am) and were dispatched to 12 laboratories in the UK and Europe in the Summer of 2017. Following analysis, using several techniques, the results were evaluated in terms of levels of precision, sensitivity and limits of detection. The results of the study showed all the radionuclides were successfully analysed by the laboratories to a level of precision and accuracy consistent with that achieved through routine analysis. The exercise demonstrated the viability and benefits of adopting the intercomparison exercise approach, which uses consensus values, to produce fit for purpose reference materials to practically address the shortage.

1 INTRODUCTION

The issue of lack of reference materials is widely acknowledged[2,3,4] and has a significant impact on the ability of radioanalytical laboratories to develop methods, produce underpinned method validation data, establish values for in-house reference materials and demonstrate ongoing proficiency. The problem is further exacerbated as laboratories extend their analytical capability to support the increasing decommissioning activities in the UK.

The ISO 17025:2017 standard[1] states the use of reference materials for ensuring correct performance of equipment (clause 6.4.1), for metrological traceability to SI (clause 6.5) and for the validation of methods (clause 7.2.2). As most UK laboratories use non standard and in-house developed methods "the validation shall be as extensive as is necessary to meet the needs of the given application or field of application".[1] The standard also requires the validity of results produced to be monitored (clause 7.7.1) on an on-going basis. The use of reference materials for all aspects is cited as one of the means of demonstrating compliance to the requirements of the standard and is practically one of the most efficient for the laboratories.

The suitability of the reference material is dependent on the scope of the analytical specification and should represent the type of samples being analysed, to underpin the analysis process including consideration of the matrix, activity range and presence of interferences both radiometric and elemental. The lack of suitable solid, mixed radionuclide materials, within the UK community is an identified area of weakness[2] with laboratories and industry expressing its concerns at several meetings, including NPL's Conference on Applied Radiation Metrology (CARM) series. Discussions have identified that the two main issues which are preventing production are the lack of availability of representative matrix material and the cost associated (estimated at £50k to £100k per material) with the production.

The UK's Analyst Informal Working Group (AIWG) conducted a survey of its members in 2015, which found that all but one of the twenty laboratories used reference materials routinely. All members expressed that the lack had an impact on their laboratories in terms of extra effort required to develop the methods or demonstrate compliance to ISO 17025. It was concluded more materials with varied radionuclide composition and matrices were required to provide the nuclear industry with the traceability it requires to underpin characterisation. The survey also ascertained the type of reference material needed. Gram quantities of material are analysed by the laboratories, which is deemed representative of a greater population, therefore the level of precision of the assigned or certified reference values should reflect this. Considering scale up and sampling uncertainties it was concluded that highly characterised, certified materials with high levels of precision were not essential, which is often one of the costly aspects of reference material production.

The ISO 17025:2017 standard[1] also states the use of inter-laboratory comparisons as a means of demonstrating validity of methods and evaluating performance. This measurement method is the lowest on the hierarchy[3] for assigning property values to materials and does not offer the same level of traceability to SI units. However, it was felt that adopting this approach could address the issues surrounding production of suitable materials. Therefore, the aim of the study was to investigate if the production of a reference material using archived materials, and adopting assigned values based on those obtained via consensus was a viable option. A laboratory had three solid materials (crustacea, mollusc and winkle) readily available, in sufficient quantities, could be easily transported to laboratories and contained a good mix of both anthropogenic and enhanced natural radionuclides (14C, 40K, 90Sr, 99Tc, 137Cs, 239+240Pu, 241Pu and 241Am). The materials were sent to thirteen laboratories with 12 returning results for each material for a range of analyses including measurement using different techniques.

2 METHODOLOGY

The characterisation of the three biota materials (crustacea, mollusc and winkle) and the assessment of the proficiency test data were undertaken following the principles outlined in ISO 13528:2015.[5] The approach the measurement results obtained by laboratories to meet the specified criteria for acceptable performance. The statistical methods for the analysis of the data, and to a lesser extent design of the exercise, has been followed as closely as possible to provide the confidence in the proposed approach.

The objective of the intercomparison exercise was to assess the performance of the participating laboratories by comparing results obtained. The limits for the proficiency test are derived from the distribution of the participants' results. This required the participants to measure the activity concentrations of the radionuclides present in the samples.

2.1 Preparation of mixed biota materials

Marine samples were collected over the last 8 years from around the Cumbrian coastline as part of the Radioactivity in Food and the Environment (RIFE) survey. Within one week of collection the samples were prepared by picking the edible fractions, oven dried at 100°C until at constant weight, ground and sieved to 1mm. Samples had been stored in airtight containers (of an appropriate size to minimise airspace).

To produce enough material for analysis the biota samples were combined according to sample category to produce three materials; crustacea (crab, lobster, nephrops), mollusc (mussels and limpets) and winkles. The material was bulked, re-sieved and thoroughly mixed. Each of the biota materials were then sub sampled to produce 14 samples for analysis that were weighed to allow correction of the results at the calculation stage for moisture uptake.

2.2 Homogeneity assessment

Four sub samples of each of the three bulked materials (crustacea, mollusc and winkles) were taken prior to the sampling and distribution of the samples to the participating laboratories. The principles of ISO 13528:2015[5] were used as a guide with the following approach adopted. Each sub sample (one from each biota material) was taken on a different day from the bulk sample that had been thoroughly mixed. Each sample was weighed to allow conversion of the weights to Bq kg^{-1} and for an assessment of moisture uptake to be made. Samples were measured on different gamma spectrometry detectors to determine the activities of ^{40}K, ^{134}Cs, ^{137}Cs and ^{241}Am.

Generally, for gamma spectrometry, measurement uncertainties of between 10-20% at the 95% confidence level are expected at the activity ranges considered. Measured uncertainties for all radionuclides except ^{60}Co (Table 1), were between 7.8% and 12%, which is below the expected measurement uncertainty. For ^{60}Co the uncertainty is higher at 19% and 26%, which reflects the lower activity concentration present in the samples and as such a greater measurement uncertainty would be anticipated. This indicates that the three materials are homogenous within the uncertainty of the technique.

Table 1 *Summary of results obtained for the gamma spectrometry analysis of the laboratory aliquots.*

Material	Geometric mean (Bq kg^{-1} dw) ± uncertainty (k=2)			
	^{40}K	^{60}Co	^{137}Cs	^{241}Am
Crustacea	281 ± 22	ND	5.07 ± 0.62	5.83 ± 0.61
Mollusc	152 ± 14	1.32 ± 0.34	8.67 ± 0.90	56.3 ± 5.3
Winkles	266 ± 22	2.25 ± 0.43	20.2 ± 1.8	73.6 ± 6.7

The ISO 13525:2015 standard[5] states that the between sample standard deviation of the replicate analysis should be below $0.3\sigma_{PT}$. This is not suitable for assessing the performance of the materials as σ_{PT} is calculated based on the robust standard deviation of the participants' data and will therefore include the uncertainty associated with sample inhomogeneity.

2.3 Distribution of intercomparison samples

Nominal 200g aliquots of each material (crustacea, mollusc or winkle) were sealed into new polythene bottles, weighed, assigned a unique identifier and then dispatched to the 13 laboratories in 4 countries. A reference date of 1st July 2017 was specified to all participants upon dispatch.

A standard reporting form was provided to all participants, which requested the following information:

- Mass of each sample;
- Contact details of the participating laboratory;
- Result for each nuclide identified (reported as Bq kg^{-1});
- Counting and method uncertainties at the k=1 level (reported as Bq kg^{-1});
- Limit of detection of each determinand (reported as Bq kg^{-1});
- Sample size analysed for each analysis;
- Standards used including yield monitor;
- Overview of method;
- Measurement technique including make and model;
- Detector type and efficiency, manufacturer, software used and calibration source details and frequency for gamma spectrometry measurements.

2.4 Review of submitted proficiency test data

All the data was assessed for obvious errors (blunder errors) with any deviant datasets rejected. The remaining data was then assessed for normal

2.5 Calculation of assigned values

The assigned values were calculated based on a consensus value for each radionuclide and material type (crustacea, mollusc and winkle) analysed from all the participating laboratories. The dataset was censored using winsorization statistics to determine the robust mean (X_{pt}), standard deviation (s* X_{pt}) and standard uncertainty (u X_{pt}) for parameter. Algorithm A (as defined in ISO13528) was used to determine the robust statistics for the dataset. The robust mean and standard deviation were calculated using the mean result of replicate data reported for each laboratory. The robust mean was taken as the assigned value for the proficiency test.

Table 2 *Assigned values for the three materials.*

Radionuclide	**Activity (Bq kg^{-1})**		
	Crustacea	Mollusc	Winkle
C14	531 ± 39	553 ± 8.4	535 ± 24
K-40	304 ± 46	142 ± 14	258 ± 29
Sr-90	0.33 ± 0.11	3.4 ± 2.0	4.9 ± 1.7
Tc-99	367 ± 60	233 ± 25	157.8 ± 7.9
Cs-137	5.40 ± 0.34	8.88 ± 0.85	20.6 ± 1.4
Pu-239+240	0.80 ± 0.10	21.4 ± 3.1	37.1 ± 5.0
Pu-241	5.2 ± 2.4	104 ± 39	167 ± 16
Am-241	6.42 ± 0.79	54.3 ± 3.5	75.4 ± 7.8

3 ASSESSMENT OF PERFORMANCE

The performance of each laboratory was assessed using the criteria in Table 3. For percentage deviation and uncertainty assessment, results were deemed as either in agreement (pass) or discrepant (fail). For all other performance parameters, data were designated as either 'pass', 'check' or 'fail, as defined in Table 3.

Table 3 *Criteria for assessment of laboratory performance.*

Criterion	Calculation	Pass criteria
Percentage deviation, *Di*	$Di = \frac{(Xi - Xpt)}{Xpt} x\, 100$	(xpt+3s*)> Di > (xpt-3s*)
Z score, z	$z = \frac{(Xi - Xpt)}{s\,*}$	|z|<2 - PASS 2 < |z| < 3 - CHECK |z| >3 - FAIL
Zeta score, ζi	$\zeta i = \frac{(Xi - Xpt)}{\sqrt{u^2(Xi) + u^2(Xpt)}}$	|ζ| <2 - PASS 2 < |ζ| < 3 - CHECK |ζ| > 3 - FAIL
Uncertainty, u(xi)	umin = u(xpt) umax = 1.5s*	umin < u(xi) < umax

4 RESULTS AND DISCUSSION

4.1 Data submission

Twelve laboratories submitted datasets that comprised of 273 individual measurements (Table 4. The only exclusions as classed as erroneous following blunder analysis) were:

- Sr-90 results for each of the materials from one laboratory;
- Cs-137 result for Crustacea.

The three materials contained a wide range of radionuclides however the results for H-3, Np-237 and Cm-244 are not considered further as the number of laboratories conducting the analysis, and hence the data set generated, was limited. The results for Pu-238 showed similar trends to those for Pu-239+240 and therefore are not presented.

4.2 Material Assessment

4.2.1 Crustacea. The crustacea material contained the lowest activities of Cs-137, Am-241, Sr-90 and Pu isotopes compared to the molluscs and winkles but the highest concentrations of Tc-99 (Table 5). The difference is due to the preferential bioaccumulation of certain radionuclides in the species. For Pu-241 analysis, the activity concentrations in the samples were close to the limit of detection with one laboratory reporting a less than value and the remaining quoting high associated uncertainties. Eighty four results were submitted for the radionuclides considered of which 88%, 95% and 97% where classed as passed for the 3 sigma deviation, z and zeta score acceptance criteria.

Table 4 *List of laboratories providing data.*

Laboratory	Country
Atkins, Hunterston B power station	UK
Cavendish Nuclear Limited	UK
Cefas Lowestoft Laboratory	UK
CRCE, Public Health England	UK
DSTL - Porton Down	UK
EPA Regional Inspectorate Dublin	Ireland
GAU-Radioanalytical	UK
Institute of Marine Research	Norway
Babcock International Group	UK
Max Rubner-Institut (MRI)	Germany
SCOTEC	UK
Wood	UK

Table 5 *Summary of results for the analysis of the Crustacea intercomparison exercise material.*

Radionuclide		Number of labs	Number of results	Within 3 sigma	Z score	Zeta score
C-14		3	5	100%	100%	100%
K-40		12	15	87%	100%	100%
Sr-90		4	7	100%	100%	100%
Tc-99	ICP	4	6	83%	100%	100%
	GFPC	1	2	100%	100%	100%
Cs-137		11	14	86%	86%	86%
Pu-239+240		6	9	89%	89%	100%
Pu-241		4	6	83%	83%	83%
Am-241	gamma	8	11	100%	100%	100%
	alpha	6	9	78%	89%	100%

4.2.2 Mollusc. For all radionuclides, except K-40, the activities in the mollusc material were at the lowest activities (Table 6). Ninety three results were submitted for the radionuclides considered of which 88%, 88% and 90% were classed as passed for the 3 sigma deviation, z and zeta score acceptance criteria similar to the statistics for the crustacea material.

Table 6 *Summary of results for the analysis of the Mollusc intercomparison exercise material.*

Radionuclide		Number of labs	Number of results	Within 3 sigma	Z score	Zeta score
C-14		3	5	100%	100%	100%
K-40		12	15	93%	93%	93%
Sr-90		5	10	70%	80%	90%
Tc-99	ICP	4	6	100%	100%	100%
	GFPC	1	2	100%	100%	100%
Cs-137		12	15	87%	93%	93%
Pu-239+240		5	10	90%	90%	90%
Pu-241		5	10	100%	100%	100%
Am-241	gamma	8	11	91%	82%	82%
	alpha	6	9	78%	78%	78%

4.2.3 Winkle. The winkle material contained the highest activities of Sr-90, Cs-137, Pu isotopes and Am-241 yet the lowest Tc-99 concentrations (Table 7). Ninety six results were submitted for the radionuclides considered of which 82%, 86% and 89% where classed as passed for the 3 sigma deviation, z and zeta score acceptance criteria. Of all three materials the winkles posed the greatest challenge to the laboratories particularly for the Sr-90, Tc-99 and Am-241 analysis.

Table 7 *Summary of results for the analysis of the Winkle intercomparison exercise material.*

Radionuclide		Number of labs	Number of results	Within 3 sigma	Z score	Zeta score
C-14		3	5	100%	100%	100%
K-40		12	15	80%	93%	93%
Sr-90		5	10	80%	70%	70%
Tc-99	ICP	4	6	50%	50%	50%
	GFPC	1	2	100%	100%	100%
Cs-137		12	15	87%	87%	93%
Pu-239+240		6	11	100%	100%	100%
Pu-241		5	10	80%	90%	90%
Am-241	gamma	8	12	67%	83%	92%
	alpha	6	9	78%	78%	78%

4.3 Radionuclide assessment

4.3.1 C-14 analysis. Three laboratories carried out the C-14 analysis on the three materials with one evaluating them in triplicate. The analysis was performed by all laboratories using the Raddec™ Pyrolyser with measurement by liquid scintillation counting. The instrumentation differed from bench top to ultra low level counters however this had little impact on the observed precision and accuracy due to the activities present.

The assigned values for the three materials indicated the activity concentrations were consistent (525 – 553 Bq kg^{-1}) with no significant variation due to sample type. All the results were within the acceptance criteria for the 3 sigma deviation, z and zeta score tests with the percentage deviations ranging from -6.0% to 18% (Figure 1).

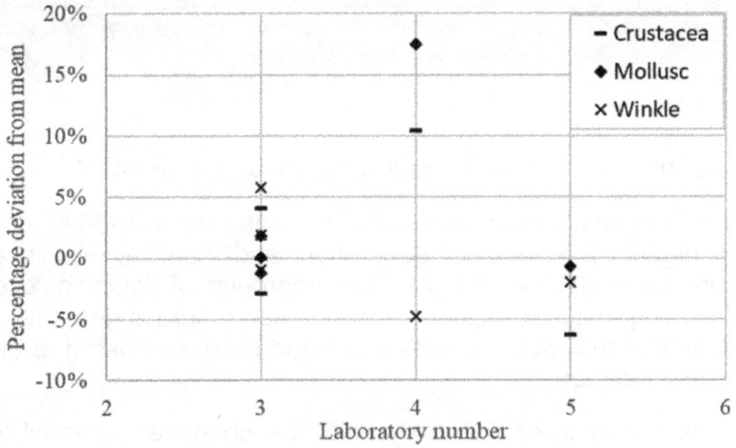

Figure 1 *Percentage deviations for C-14 results of the three intercomparison exercise materials.*

4.3.2 K-40. For the crustacea and mollusc materials the results were similar with more than 87% classed as a pass for the three statistical criteria (deviation, z and zeta score). For each sample type the results that were outside the acceptance criteria were from two laboratories who did not routinely analyse for K-40 and also had a significant delay between receipt of samples and measurement.

The variation in K-40 measured for the analysis of the winkle material was similar (Figure 2) with 80% of the laboratories producing results within the three sigma limit and 93% within the zeta and z score acceptance criteria. Deviations ranged from and deviations ranging from -29% to 21%. The assigned values were comparable to those of crustacea and therefore the variation is not due to being near the limit of detection.

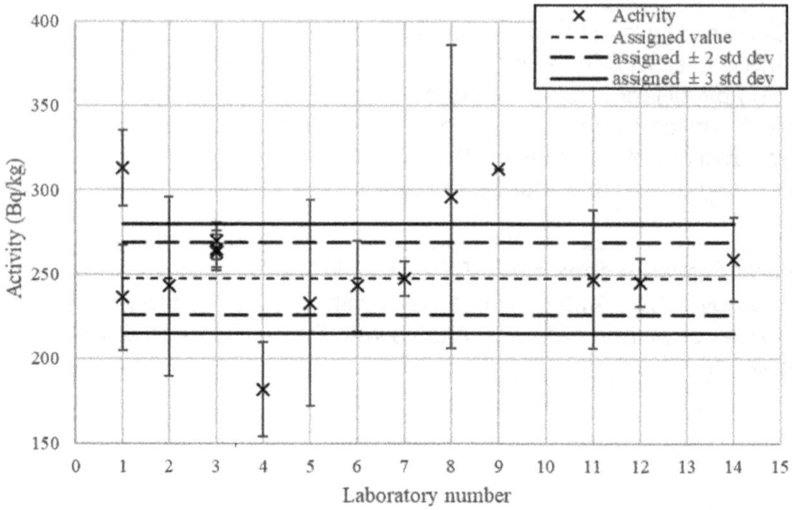

Figure 2 *Results for the analysis of K-40 in the winkle material.*

4.3.3 Sr-90. Five laboratories carried out the analysis of Sr-90 by carrying out an acidic digest, followed by separation and purification via extraction chromatography whilst two adopted the approach of liquid-liquid extraction followed by a sequential precipitations. The latter approach generally produced limits of detection that were at least a five fold lower compared to the column chemistry methodology.

All the laboratories for the analysis of the crustacea material, produced results within the three assessment criteria despite being at the lowest activity. The number of results submitted was lower as some reported the activities as below the limit of detection. The laboratories reported high uncertainties of between 17% and 73% with the results close to the limit of detection (ranging from 0.024 Bq kg^{-1} to 0.23 Bq kg^{-1}).

For the mollusc and winkle material 80% and 90% of the laboratories respectively were classed as a pass for the three criteria. However a significant level of variation was observed (Figure 3) especially for the mollusc material where the assigned value uncertainty was at 60% (at 95% confidence limit). Laboratory one analysed the samples in quadruplicate with the results showing the same trends as the wider data set and also a positive bias compared to the assigned values. As the analysis was carried out on using identical parameters (timescales, analysts, sample preparation including digestion, methodology, instrumentation) the source of the variation is unclear and could be due to inhomogeneity in the samples submitted to the laboratories. This was not observed when the samples were analysed by gamma spectrometry however the sample size was much larger (~200g compared to 10g) and the variation could be on a smaller scale.

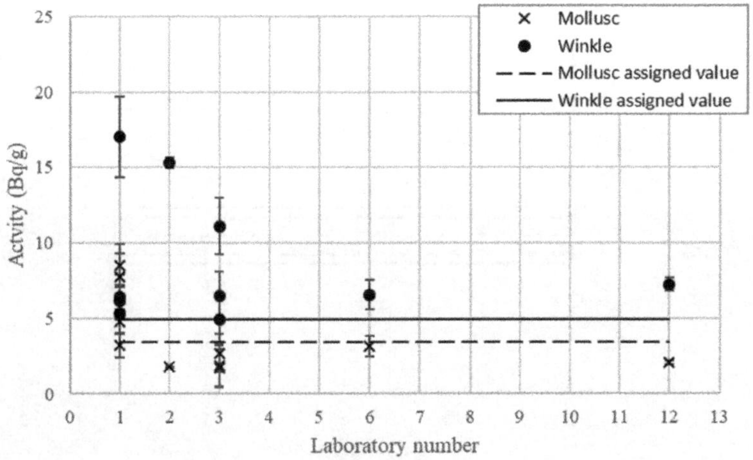

Figure 3 *Comparison of Sr-90 values for mollusc and winkle materials.*

4.3.4 Tc-99. Technetium-99 analysis was carried out by 4 labs with all using Inductively Coupled Plasma Mass Spectrometry (ICP-MS) for the final determination and one also carrying the measurement out using Gas Flow Proportional Counting (GFPC). For all three materials the results obtained using GFPC were within all three acceptance criteria. This was the same for the analysis of the mollusc material and except one criteria for the analysis of the crustacea by ICP-MS.

For the winkles half of the samples were outside the acceptance criteria for ICP-MS measurement. Although the activity concentration in the winkle was the lowest the concentrations are significantly above the limit of detection. It is more likely, given that the remainder of the results are consistent (Figure 4), that the variability is due to an analytical issue originating from one laboratory that submitted results using two measurement techniques. It should be noted that the sample digestion methodologies, as well as the instrumentation, differed with the samples for ICP-MS analysis were digested in nitric acid in a microwave whereas those for proportional counting were ashed in a furnace and dissolved in HCl with addition of hydrogen peroxide. The digestion procedure was not stated by labs 2 and 6 in detail whilst lab 5 carried out acid digestion post ashing. The impact of different digestion techniques can not be assessed based on the limited data set but it does identify the need for laboratories to consider prior to analysis.

4.3.5 Cs-137. As with K-40, Cs-137 is easily analysed via gamma spectrometry and a high number of results submitted. The activity concentrations in the three materials ranged from 5.39 Bq kg^{-1} to 20.6 Bq kg^{-1} and were consistent with over 85% of the dataset within the three acceptance criteria. High uncertainties of ~50% were quoted for the results submitted by laboratory 8.

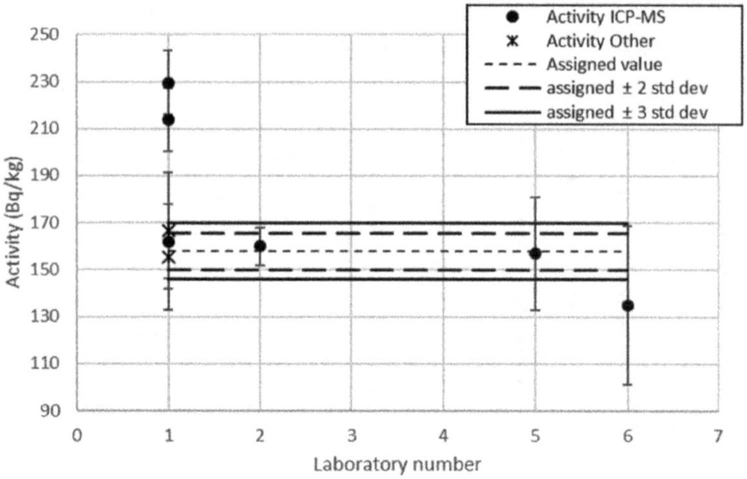

Figure 4 *Results for the analysis of Tc-99 in winkle samples.*

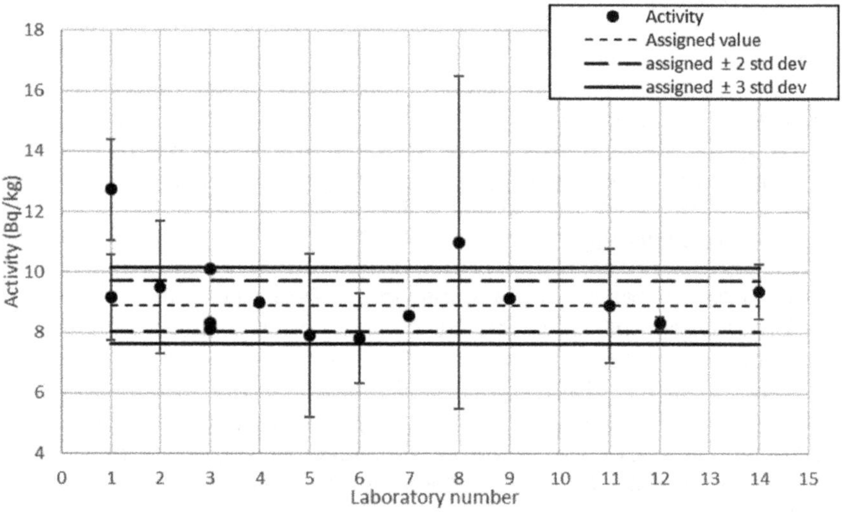

Figure 5 *Analysis of Cs-137 in molluscs.*

4.3.6 Pu isotopes. Similar to the results obtained for Cs-137 the dataset showed good agreement across the laboratories for the analysis of the three materials. Over 85% of the results were within the three acceptance criteria with all the results for Pu-241 in molluscs and Pu-239+240 in winkle classed as a pass. Generally the variations for the Pu isotopes (Pu-239+240 and Pu-241) were consistent (Figure 6) as illustrated by the percentage deviations, which is to be expected as the Pu-241 analysis is carried out on the Pu alpha spectrometry source.

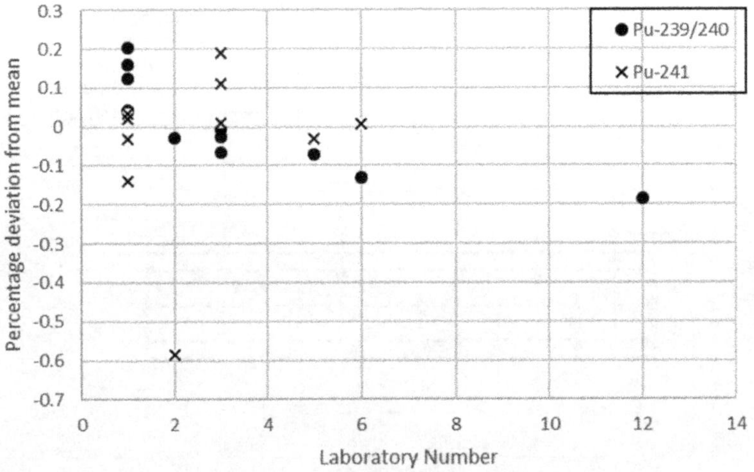

Figure 6 *Pu-239+240 and Pu-241 results in winkles.*

The crustacea samples showed the highest variation in dataset which was probably due to the lowest activities presenting the samples (Pu-239+240 at 0.81 Bq kg-1 and Pu-241 at 5.28 Bq kg-1) which were near to some of the limits of detection quoted especially for Pu-241. However, the variability is at a level expected for the analysis of this type across 12 laboratories, generating 56 results.

4.3.7 Am-241. The analysis of Am-241 was carried out by 12 laboratories with 3 using alpha spectrometry for the final measurement, 5 using gamma spectroscopy and 3 reporting data using both techniques. For the crustacea material, which contained the lowest activity concentration of Am-241 (6.42 Bq kg^{-1}) all the gamma spectrometry results and 87% of the alpha spectrometry results were within the three acceptance criteria (Figure 7). A ranked plot of all the crustacea data indicated a symmetrical, unimodal distribution consistent with a normal distribution of data (Figure 8). With no significant differentiation between the results obtained by the different methodologies including digestion procedures and measurement techniques.

The mollusc material contained over 8 times the activity of Am-241 compared to the crustacea. For the analysis of the mollusc material 82% of the gamma spectrometry derived results and 78% of the alpha spectrometry were within the three acceptance criteria. For the winkle material, (74.2 Bq kg^{-1}) the number of results that were within the 3 sigma deviation when analysed using gamma spectrometry was 58% compared to 78% by alpha spectrometry. Although a material has higher activities the number of laboratories within the acceptance criteria is lower. In comparison to the crustacea the assigned value uncertainties are lower, largely due to the higher activities and less variation across the measurements. However this generates tighter tolerance levels and can lead to a greater number of outliers.

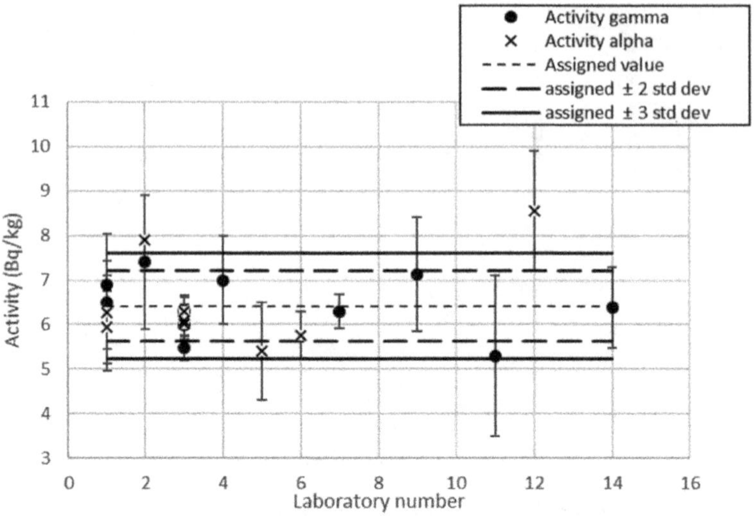

Figure 7 *Results for the analysis of Am-241 in crustacean.*

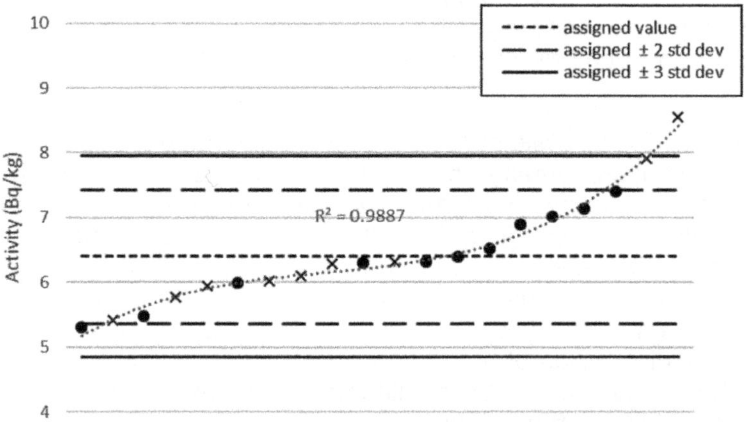

Figure 8 *Ranked distribution of Am-241 results in crustacea. × indicates alpha spectrometry and • indicates gamma spectrometry measurements.*

5 SUMMARY OF RESULTS

All the laboratories undertook the analysis of the three materials for the gamma emitting radionuclides (K-40, Cs-137 and Am-241) however less than half submitted results for analysis that required more involved analysis. As a result the data set for some of the radionuclides (e.g. C-14 and Tc-99) was small limiting the interpretation of data and prevented the statistical analysis of the H-3, Cm-244 and Np-237 activities that were present in all three materials.

The uncertainties associated with the assigned values were as expected ranging from 2% to 16% for most radionuclides. The exception was Sr-90 in all three materials and Pu-241 in the crustacea and mollusc samples, which ranged from 33% to 60%. The increased variation was most likely due to the activity levels present being close to the limit of detection for the methods and the relatively small data set may not have limited the identification of outliers being recognised.

The analysis of Sr-90 and Tc-99 using differing methods and measurement techniques identified differences in the data sets. Due to the limited number of results the root cause can not be determined but does identify the need to carefully consider digestion techniques and the homogeneity of the materials being used for reference on a macro (200 g) and micro (1-5g scale) sample.

However, the results do show the viability of adopting the intercomparison exercise approach. The cost associated of conducting the study does not significantly fall on one laboratory to carry out the characterisation of the materials to a high level of precision as the assigned values are determined by consensus. The materials are suitable for use to develop and validate methods and to evaluate performance. The economy of this approach must be balanced by the need of the participants and the level of quality required of the result (e.g. precision and accuracy).

6 CONCLUSIONS

Three mixed biota QC materials were successfully produced from surplus laboratory samples providing solid matrices that contained a range of alpha, beta and gamma radionuclides. The interlaboratory comparison exercise, lead to the comprehensive characterisation of the materials that significantly reduced the financial burden to any one organisation.

The results showed good agreement being observed between all participating laboratories across a range of analytical techniques. The level of precision and accuracy quoted for the assigned values reflect those which are generated by the laboratories and therefore deemed suitable.

The variation in the Tc-99 has identified that the impact of utilising differing sample digestion and instrumentation. The results, though not conclusive, illustrate if the composition is unknown it is advisable to consider digestion technique.

The analysis of the three materials has demonstrated the benefits of using "surplus" material and adopting an intercomparison exercise approach to determine consensus values to produce fit for purpose reference materials that practically addresses the shortage that exists.

CONFLICTS OF INTEREST

There are no conflicts to declare.

ACKNOWLEDGEMENTS

The authors particularly thank all participants for their valuable contribution to the intercomparison exercise. We would also like to thank Paul Blowers (Cefas) for supplying the three materials derived from historic samples used in the RIFE project. In addition thanks to Tractebel for funding the time required to present and author this paper.

NOTES AND REFERENCES

1. ISO/IEC 17025:2017, General requirements for the competence of testing and calibration laboratories, 2017.
2. S. J. Parry, Quality assurance in the nuclear sector, *Radiochim Acta*, 2012, **100**, 495–501.
3. European Accreditation, *The Selection and Use of Reference Materials*, 2003, EA-4/14 INF:2003.
4. P. Warwick, I. Croudace and R. Mason, Maintaining radioanalytical data quality: The challenges of decommissioning analysis, *Nuclear Futures*, 2011, **7**(55), 44–53.
5. ISO 13528:2015, Statistical methods for use in proficiency testing by interlaboratory comparison, 2015.

DETERMINATION OF NATURAL RADIOACTIVITY IN IRON AND STEEL MATERIALS

Franck Dal-Molin[a,b*], Adam Sutcliffe[a], David R. Anderson[a] and David Read[b,c]

[a]Tata Steel Group Health, Safety & Environment, Swinden Technology Centre, Rotherham, S60 3AR, UK; [b]Department of Chemistry, University of Surrey, Guildford, GU2 7XH, UK; [c]National Physical Laboratory, Hampton Road, Teddington, Middlesex, TW11 0LW, UK

*franck.dal-molin@tatasteeleurope.com

ABSTRACT

It has been known since the 1990s that two natural radioisotopes from the uranium-238 (^{238}U) decay series, polonium-210 (^{210}Po) and lead-210 (^{210}Pb), originally present in trace amounts in raw materials, are volatilised and concentrate in the form of dusts during iron ore sintering. In the UK, most of the dust generated during this process is collected by means of electrostatic precipitators and recycled back into the production system using conveyor belts. Nevertheless, a small proportion passes into the atmosphere via stack emissions and some fugitive dusts can also escape into the workplace during maintenance operations. Tata Steel UK Ltd, a major European steel making company, has developed and validated in-house radioanalytical methods for the measurement of ^{210}Po and ^{210}Pb in a wide range of iron-making materials including raw feedstock, waste dusts, occupational and emission filter samples. The data gathered have enabled a better understanding of the fate of ^{210}Po and ^{210}Pb throughout the integrated steel making route, providing essential information to support environmental permits for discharges to the atmosphere and for confirming that chronic exposure to these two natural radioisotopes does not lead to significant radiological doses to the workforce. Additionally, since the implementation of the BSS Directive 2013/59 Euratom and the Construction Products Regulation (CPR), there is a need for the European steel industry to characterise the levels of radium-226 (^{226}Ra), thorium-232 (^{232}Th)

and potassium-40 (^{40}K) in slag materials and confirm that those materials do not pose a significant risk of internal and external exposure to radiation when reused or recycled in building materials. This paper highlights the technical challenges encountered when measuring those natural radioisotopes in various iron-making materials, including the difficulty of validating radioanalytical methods in the absence of suitable certified reference materials.

1 THE INTEGRATED STEELMAKING ROUTE

The integrated steelmaking route comprises of separate compartments to prepare the raw materials into the suitable forms required to convert iron ore into semi-finished steels and finished steel products (Figure 1).

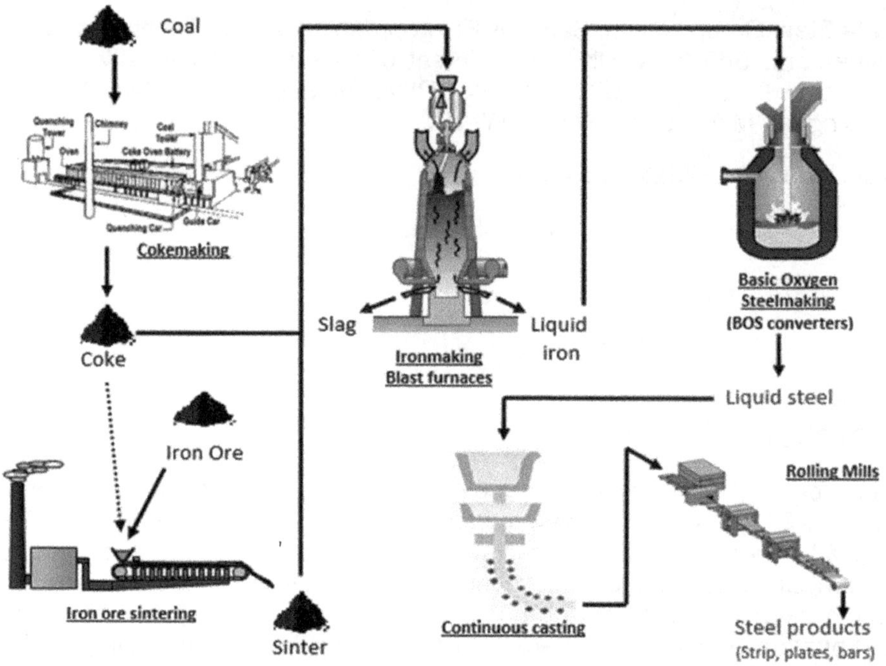

Figure 1 *The integrated steelmaking route.*

The principal purpose of iron ore sintering plants is to convert fine iron ores into a clinker-like agglomerate for use in modern high performance blast furnaces. The sintering process also plays an important role in the recycling of ferroginous materials arising from the downstream processes in steel production. The raw materials used in the sintering process are blended in the necessary portions into a sinter mix comprising of high-grade iron ores, fine coke, limestone, recycled waste gas dusts from the process (captured by electrostatic precipitators (ESP)) and various residues such as the dust and slurries collected from the blast furnace gas cleaning plant. The iron ore mix typically constitutes 65 to 80% of the sinter blend and, therefore, the iron ores have a major influence on the overall chemical composition of the blend.

The finished sinter represents a major part of the burden of blast furnaces, where primary reduction of ore oxides takes place to produce liquid iron. The charging system at the top of the furnace acts as a valve to prevent escape of gas that is taken to the gas cleaning plant where it is cleaned to allow its reuse as a fuel across the works.[1] It contains a large amount of dust that is largely removed through a dry first step of the gas treatment system either by using a dust catcher or a dry cyclone process. The remainder is then scrubbed from the gas by means of wet scrubbing to produce slurries. Pulverised and granulated coal is injected into the blast furnace to provide additional heat and to reduce coke requirements. Liquid iron and slag are removed by continuous tapping from the hearth and the liquid iron is taken to the BOS plant to produce liquid steel. The steel is then cast into semi-finished material ready for use.

2 DEVELOPMENT OF NOVEL RADIOANALYTICAL TECHNIQUES

2.1 Determination of ^{210}Po in ironmaking materials and stack emission samples by alpha spectrometry

Stack emission filters, raw materials and steelmaking dusts were spiked with a known amount of certified polonium-209 (^{209}Po) solution (R33-01, National Physical Laboratory (NPL), UK) and digested in a microwave system (Mars 5, CEM, USA). Stack emission samples were digested using 10 cm^3 of a mixture of hydrochloric acid (HCl 37%) and nitric acid (HNO_3 69%) (1:4 ratio). Other materials were digested using 12 cm^3 of *Aqua Regia* (HCl: HNO_3; 3:1). All samples were then filtered and the solutions evaporated to dryness and diluted in 30 cm^3 of 10 M HCl. Iron (III), a major interference for the analysis of ^{210}Po by alpha spectrometry,[2] was extracted from the concentrated hydrochloric acid solution using liquid-liquid extraction with di-isopropylether. After extraction, the concentrated hydrochloric acid solutions were evaporated to dryness a second time and taken up in 100 cm^3 of 0.5 M HCl. A silver disc (ESF005, CooksonGold, UK) was then placed inside each solution for 6 hours at 80-90 °C to allow the deposition of ^{210}Po and ^{209}Po. Each disc was counted by alpha spectrometry (Octête Plus Alpha-Spectrometer and Alpha Ensemble 8 systems, ORTEC, USA). Quantification of ^{210}Po was achieved by comparison with the results for a known amount of ^{209}Po yield monitor and the spectrometers were calibrated using two certified radioactive sources (NK220 and AMR21, Isotrak, UK).

2.2 Determination of ^{210}Pb in stack emission filters and raw materials by alpha spectrometry

As it is not possible to directly determine ^{210}Pb accurately at low activities by gamma spectrometry for stack emission filters and raw materials, the solutions from ^{210}Po analysis were kept for a minimum of three months to allow for ^{210}Po in-growth from the decay of ^{210}Pb. Polonium-210 is deposited on a silver disc, as described in the previous section and this enabled back-calculation of ^{210}Pb in the sample.

2.3 Determination of ^{210}Pb in sinter and blast furnace dusts by gamma spectrometry

Dust samples from the sintering and blast furnace processes were counted on a well-type Coaxial P-type High Purity Germanium detector (57.8 mm active diameter). Spectra were acquired and counts were recorded using Maestro-32 software (ORTEC, USA). Information concerning the chemical composition and bulk density were integrated into a correction model that compares count rates from the target (unknown) sample with those obtained from an in-house reference material made of ultrapure hematite powder (Alfa Aesar, UK), and spiked with a known amount of ^{210}Pb using a certified reference solution (R22-02, National Physical Laboratory, UK). Each sample, including the reference hematite, was analysed in uniform sample holders of 5 cm diameter, at a fixed position to the detector and using masses ranging from 9.9 to 10.1 g to minimise the geometry effects.[3]

2.4 Determination of ^{226}Ra in iron ores by alpha spectrometry

Iron ore materials were digested in duplicate in a microwave system (Mars 5, CEM, USA) using 12 cm^3 of *Aqua Regia* (HCl: HNO3; 3:1). Each sample digest was then transferred to a 50 cm^3 graduated Falcon tube, diluted to 30 cm^3 using deionised water and finally spiked with approximately 1 Bq of a certified solution of radium-223 (^{223}Ra) (NPL, UK). Radium-223 is generally used in radiopharmaceutical applications to treat metastatic cancers in bone.[4] Although its relatively short half-life (11.4 days) is a limiting factor for its use as a tracer in alpha spectrometric applications, it allows the direct determination of chemical yield by alpha spectrometry. In order to, extract iron from the solutions from iron ore digestions, 10 cm^3 of concentrated ammonia was added. The solutions were then centrifuged for 2.5 minutes at 6,000 rpm (Z206A, Hermle, Germany) and the supernatant passed through a filter paper (541, Whatman, UK). The duplicate samples were recombined in a beaker and the solutions were acidified using 5 cm^3 of acetic acid to ensure an acidic pH. Then, 3 g of ammonium sulphate ($(NH_4)_2SO_4$) and 0.1 cm^3 of a solution of barium chloride ($BaCl_2$) at 0.75 $\mu g/cm^3$ were added to initiate the precipitation of $Ra(Ba)SO_4$. Finally the solutions were stirred manually to ensure complete dissolution of $(NH_4)_2SO_4$ and were transferred to funnel units pre-packed with Resolve® filters of 0.1 µm porosity and 25 mm diameter (Eichrom, USA), pre-conditioned with 5 cm^3 of ethanol (80%) (see Figure 2). Resolve® filters are specifically designed to prepare alpha sources uniformly and maximise analyte recovery onto the filter.

After eluting the entire solution, the precipitates collected on the filters were dried under a heat lamp for 1 hour and mounted onto a stainless steel disc of 25 mm diameter (Triskem International, France). Each disc was counted by alpha spectrometry (Octête Plus Alpha-Spectrometer and Alpha Ensemble 8 systems, ORTEC, USA). Quantification of ^{226}Ra was achieved by comparison with the results for a known amount of polonium-215 (^{215}Po), one of the decay products of ^{223}Ra.

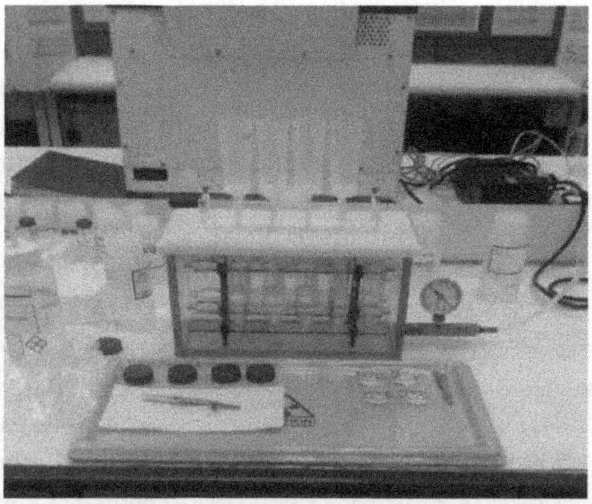

Figure 2 *Photograph of Eichrom Resolve® filters prepacked in disposable funnel units.*

2.5 Determination of ^{226}Ra in blast furnace slag materials by gamma spectrometry

Approximately 10 g of blast furnace slag materials were first sealed in a plastic scintillation vial of 2.1 cm diameter to prevent radon gas escape and allow equilibration between bismuth-214 (^{214}Bi), lead-214 (^{214}Pb) and radon-222 (^{222}Rn). After a month, the still-sealed samples were directly counted using the same HPGe detector and acquisition software described in 3.3. In contrast to the ^{210}Pb gamma spectrometry method, the certified reference material (CRM) IAEA-434 phosphogypsum was used as a reference material and the two main peaks of ^{214}Pb and ^{214}Bi, at 352 and 609 keV respectively, were used to indirectly determine the ^{226}Ra activity concentration ranges present in blast furnace slag samples. The CRM IAEA-434 was collected from a processing plant in Gdansk (Poland) in 2003.[5] Its main components are calcium sulphate ($CaSO_4$), phosphorous pentoxide (P_2O_5), iron, silicon dioxide (SiO_2) and aluminium oxide (Al_2O_3), which presents some similarities with the chemical composition of a typical blast furnace slag material. Although the photoelectric absorption process is not as dominant at energies above 200 keV, the bulk chemical composition information and the bulk density information were integrated into a correction model following the same approach described in section 3.3.

3 RESULTS AND DISCUSSION

3.1 Validation methodologies

In the last twenty years, the UK steel industry has been legally permitted to discharge ^{210}Po and ^{210}Pb into the environment via its sinter plant stacks. In order to support its compliance with this specific environmental permit and estimate the radiological risk of NORM residues better, the radioanalytical methods associated with the measurement of these two radioisotopes were validated and accredited according to the EN IEC/ISO 17025:2005.[6] As there are currently no suitable and traceable certified reference materials available, several in-house reference materials (RMs) were developed. One iron ore and one sinter dust were selected to be developed as RMs for the alpha spectrometry method (section 3.1) and the same sinter dust along with a blast furnace dust were selected for the gamma spectrometry method presented in section 3.3. When analysing stack emission samples, one blank filter is routinely spiked with a known amount of a certified ^{210}Pb standard solution (R22-02, NPL, UK) and used as a reference for result validation (section 3.2). Further studies were needed to confirm the suitability of these in-house steelmaking materials as CRMs. Between 2010 and 2015, two separate Round Robin exercises were also initiated in partnership with five external UK radioanlytical laboratories in order to support the potential future certification of these materials. Meanwhile, the two CRMs - IAEA-434 phosphogypsum and IAEA-384 marine sediment - have been routinely used to demonstrate the robustness of the proposed methods. The results from the monitoring of ^{210}Po in the CRM IAEA-434 between May 2013 and February 2016 are presented in Figure 3.

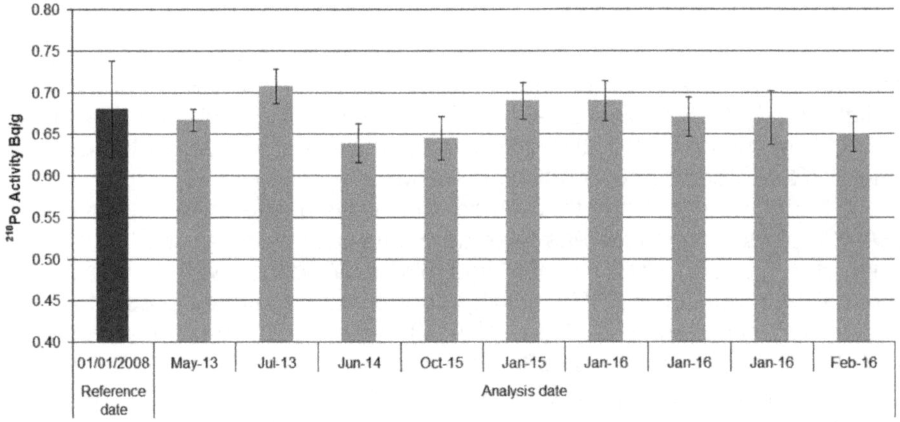

Absolute uncertainty k = 2

Figure 3 *Comparison between the IAEA-434 certified 210Po reference activity concentration and in-house values obtained between May 2013 and February 2016.*

Where possible, the results from two different analytical routes were also used as a validation mean. For example, a series of experiments was undertaken to compare ^{210}Pb results obtained from gamma spectrometry with those obtained from alpha spectrometry via ^{210}Po in-growth. In this study, eleven waste gas dust materials from the two UK sinter plants were collected in 2013 and analysed using both alpha and gamma spectrometry methods. The results are presented in Table 1 and Figure 4. As may be seen from Table 1, the results obtained by alpha and gamma spectrometry were in very good agreement for ^{210}Pb with typically less than 8% difference between the methods. Although the estimated uncertainties using the in-growth method were higher, mainly due to the potential presence of residual ^{210}Po before in-growth, the results obtained were very comparable with the gamma spectrometry results and thus reinforce the validity of the results obtained from the two methods.

Table 1 *Comparison of ^{210}Pb activities in a series of waste gas dust materials from both UK sinter plants using alpha and gamma spectrometry.*

Plant	Sample	^{210}Pb by gamma spectrometry (Bq/g)	^{210}Pb by alpha spectrometry (Bq/g)	Absolute Difference (%)
A	1	1.98 ± 0.13	2.03 ± 0.57	3
	2	3.12 ± 0.21	2.91 ± 0.81	7
	3	2.45 ± 0.15	2.30 ± 0.68	6
	4	2.31 ± 0.15	2.12 ± 0.63	8
	5	2.94 ± 0.19	2.80 ± 0.76	5
	6	2.67 ± 0.17	2.49 ± 0.74	7
B	1	5.67 ± 0.38	5.86 ± 1.78	3
	2	5.96 ± 0.37	5.86 ± 1.70	2
	3	5.96 ± 0.38	5.80 ± 1.73	3
	4	5.89 ± 0.42	5.88 ± 1.66	<1
	5	4.21 ± 0.27	4.17 ± 1.32	1

Absolute uncertainty k = 2

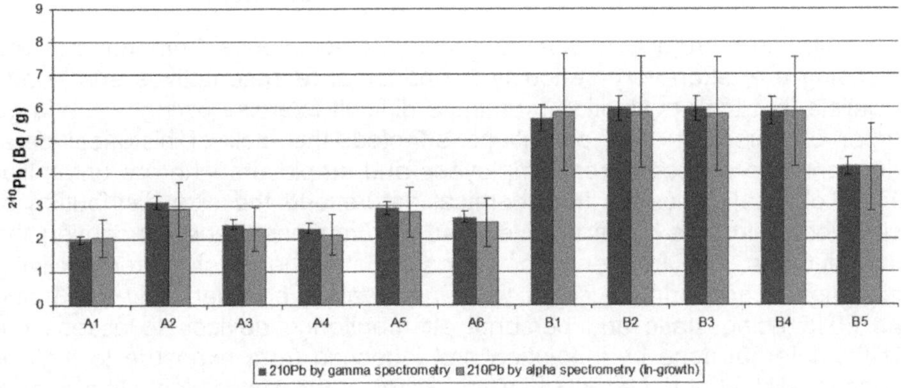

Absolute uncertainty k = 2

Figure 4 *Comparison of ^{210}Pb activities in a series of waste gas dust materials from both UK sinter plants using alpha and gamma spectrometry.*

Finally, knowing that iron ore materials are in secular equilibrium, the ^{226}Ra results obtained by alpha spectrometry could also be compared directly with the accredited ^{210}Po and ^{210}Pb results. The results from this study carried out in 2017 are presented in Figure 5.

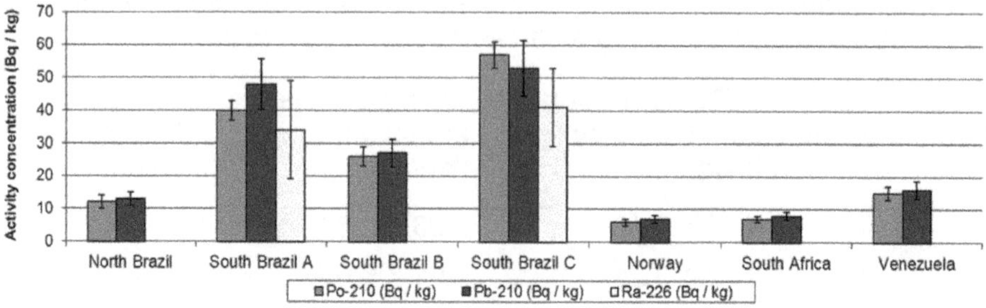

Absolute uncertainty k = 2

Figure 5 *Comparison between ^{226}Ra, ^{210}Pb and ^{210}Po activity concentrations in different iron ores.*

As can be seen in Figure 5, only two ^{226}Ra results could be obtained. Nevertheless, the two results were found to be in a relatively acceptable agreement with their respective ^{210}Po and ^{210}Pb activity concentrations. High background levels originating from residual contamination could be observed in some alpha counting chambers. As a consequence, the results for the other five studied iron ores, which presented lower natural activity concentrations, could not be statistically exploited as being lower than their respective limits of detection. Nevertheless, the ^{210}Pb results obtained from ^{210}Po in-growth seemed to be in good agreement with the ^{210}Po results obtained by alpha spectrometry, which helped validating the two methods.

3.2 Chronic exposure and radiological dose assessment

The health risks from exposure to natural radioisotopes from mineral ore processing are taken very seriously by industrial representatives and health specialists, but their evaluation remains a difficult exercise owing to the large number of personnel and operations affected, the lack of historical data, understandable concerns from employees and employers who are unfamiliar with radiological concepts, the chemical nature and the size distribution of occupational airborne dusts involved and assumptions made concerning the actual duration, the frequency of exposure and the level of radioactivity. Occupational monitoring surveys were undertaken between July 2013 and June 2015 using static and personal air monitoring devices to assess the potential internal dose by inhalation and ingestion from exposure to fugitive occupational dust enriched in ^{210}Pb and ^{210}Po at the sinter plant. The annual internal effective doses were then calculated in line with standard methodologies. Information on the characteristics of the dusts and aerosols together with *in-vitro* solubility experiments were used to select the most appropriate dose coefficients from the latest ICRP recommendations.

For example, dissolution tests were performed on occupational particulate matter 10 µm or less in diameter (PM_{10}) in a synthetic Gamble's lung fluid to verify the suitability of the default moderate solubility level for ^{210}Po and fast solubility level for ^{210}Pb from ICRP 119.[7] As can be seen in Figure 6, the dissolution of ^{210}Po in a respirable dust sample seemed to increase slowly to around 1.3% after six months, which corresponded to a slow dissolution rate (Type S material). The results from ^{210}Po in-growth in all aliquots were found to be below the limit of detection of the alpha spectrometry method, also suggesting slow dissolution of ^{210}Pb in lung fluid. Thus, the use of the standard inhalation types for both ^{210}Pb and ^{210}Po and default dose coefficients presented in Table 2 will underestimate the effective doses received in this specific exposure context. In a public exposure context, ICRP Publication 119 provides committed effective dose coefficients for Type S material of 5.6 x 10^{-6} $Sv.Bq^{-1}$ for ^{210}Pb and 4.3 x 10^{-6} $Sv.Bq^{-1}$ for ^{210}Po. These dose coefficients were used to calculate the potential dose figures for the different studied workplaces.[8]

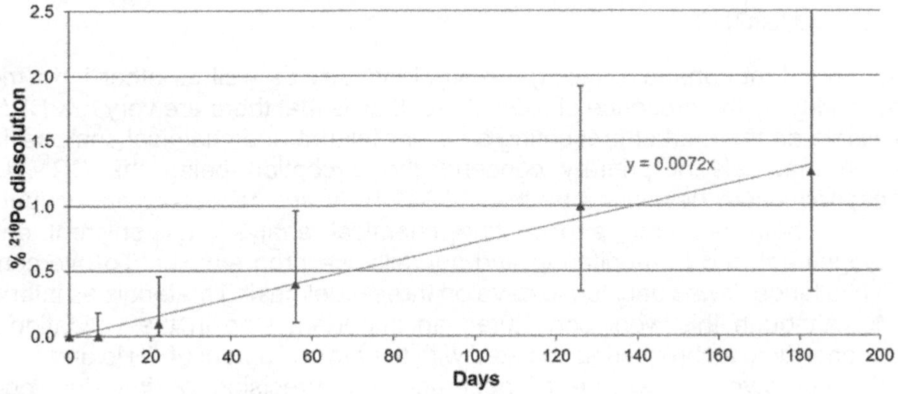

Figure 6 *Dissolution rate of ^{210}Po in occupational PM_{10} in a synthetic Gamble's lung fluid solution.*

Table 2 *Effective dose coefficients for intakes of ^{210}Po and ^{210}Pb via inhalation (E_{inh}) and ingestion (E_{ing}).*[7]

Nuclide	t½	E_{inh} ($Sv.Bq^{-1}$)				E_{ing} ($Sv.Bq^{-1}$)	
		Type	f1	1 µm AMAD	5 µm AMAD	f1	
^{210}Po	138 days	F	0.1	6.0 x 10^{-7}	7.1 x 10^{-7}	0.1	2.4 x 10^{-7}
		M	0.1	3.0 x 10^{-6}	2.2 x 10^{-6}		
^{210}Pb	22.3 years	F	0.2	8.9 x 10^{-7}	1.1 x 10^{-6}	0.2	6.8 x 10^{-7}

3.3 Recycling of NORM residues in building materials

Another regulatory challenge linked with the production of NORM residues is related to specific requirements for their use and recycling in building materials, particularly within the European Union following the implementation of the BSS Directive 2013/59 Euratom[9] and the Construction Products Regulation (CPR).[10] Natural radioactivity is present in all building materials of geological origin. However, some building materials may be composed of industrial by-products or residues known to contain enhanced levels of natural isotopes, such as red mud from the aluminium industry, ashes from the coal industry or blast furnace slag materials from the steel industry.[11] Indeed, granulated blast furnace slag is used worldwide to make durable concrete structures in combination with ordinary cement. Its two major uses are in the production of quality-improved slag cement, namely Portland blast furnace cement (PBFC) and high slag blast furnace cement (HSBFC), with content ranging typically from 30 to 70%; as well as in the production of ready-mixed or site-batched durable concrete. The levels of ^{226}Ra in several blast furnace slag materials were investigated by gamma spectrometry and were all below 0.25 Bq/g.

4 CONCLUSIONS

One important constraint facing the steel industry as well as other industries challenged by the presence of NORM residues is that there are very few CRMs available on the market presenting a similar texture and chemical composition to the materials of primary concern; the exception being the CRM-434 phosphogypsum developed by the IAEA. CRMs are necessary and valuable tools in both research and routine chemical analysis for environmental management and in monitoring and surveillance programmes. To overcome this challenge, it was decided to develop three steelmaking materials as internal RMs. Although this work constituted an important step in the validation of radioanalytical techniques associated with the measurement of ^{210}Po and ^{210}Pb, it did not give a measure of accuracy and precision of the developed methodologies. Further studies are necessary to confirm the suitability of these RMs as CRMs and this work would be highly beneficial for the steel industry worldwide.

The main uncertainty for estimating internal doses in NORM impacted industrial environments is related to the determination of an intake from chronic exposure. However, another important source of uncertainty arises from the use of biokinetic and dosimetric models. The retention characteristics of ^{210}Po and ^{210}Pb in the body were originally evaluated to select the most suitable default coefficients for converting intakes into dose. The results from *in vitro* dissolution tests of PM_{10} using a synthetic lung fluid and from the analysis of their size distribution in the workplace revealed an important risk of underestimating doses by inhalation from using the default inhalation dose coefficients for both ^{210}Po and ^{210}Pb in a working exposure context. Thousands of workers are chronically exposed to industrial NORM around the world in a wide range of activities involving the presence of relatively dry, solid particulate materials which can potentially lead to more significant doses. In the majority of cases, radium isotopes are of principal concern rather than ^{210}Po and ^{210}Pb. It is known that chronic exposure to low doses of ionising radiation increases the risk of death from solid cancers. There is therefore an urgent need to develop new ingestion and inhalation dose coefficients for natural radioisotopes to improve the estimation of health risks from working exposure to industrial NORM.

Potential radiation exposure from building materials is mainly caused by direct gamma radiation from the presence of ^{226}Ra, ^{232}Th and ^{40}K. As more than one radionuclide could contribute to the external dose, it was practical for the regulators to present an investigation level in the form of an activity concentration index to identify potential building materials that may lead to an additional significant public indoor exposure to gamma radiation taking into account the ways and quantities in which materials are used in building. In this study, the level of ^{226}Ra in blast furnace slag materials was investigated but not the levels of the other two natural radioisotopes. Further radioanalytical development work would be required to define the radiological index for typical blast furnace slag materials and to confirm that those materials do not pose a significant risk of external exposure to radiation when reused or recycled in building materials.

REFERENCES

1. European Commission, Integrated Pollution Prevention and Control (IPPC), Best Available Techniques for the Production of Iron and Steel, 2012.
2. K. M. Matthews, C.-K. Kim and P. Martin, Determination of ^{210}Po in Environmental Materials: A review of analytical methodology, *Appl. Radiat. Isot.*, 2007, **65**, 267–279
3. F. Dal Molin, D. R. Anderson and D. Read, Determination of polonium-210 and lead-210 in iron- and steel-making materials, in *Environmental Radiochemical Analysis V*, 2015, pp. 175–184
4. R. H. Larsen and G. Henriksen, The preparation and use of radium-223 to target calcified tissues for pain palliation, bone cancer therapy, and bone surface conditioning, *Patent CA 2358498 C*, 2008.
5. IAEA, Reference material IAEA 434 naturally occurring radionuclides in phosphogypsum, *Analytical Quality in Nuclear Applications* AQ/17, 2010.
6. International Standard ISO/IEC 17025:2005 General Requirements for the Competence of Testing and Calibration Laboratories.
7. ICRP, Compendium of dose coefficients based on ICRP Publication 60, Publication 119, 2012, *Ann. ICRP*, **41**(Suppl.).
8. F. Dal Molin, D. R. Anderson, R. Fisher, D. Frost and D. Read, Dose assessment from chronic exposure to industrial NORM in iron ore processing, *J. Radiol. Prot.*, 2017, **37**, 402–421.
9. European Commission, Council Directive 2013/59/Euratom of 5 December 2013 laying down basic safety standards for protection against the dangers arising from exposure to ionising radiation, and repealing Directives

89/618/Euratom, 90/641/Euratom, 96/29/Euratom, 97/43/Euratom and 2003/122/Euratom, *OJ* L13, 2014.
10. European Commission, Regulation No. 305/2011 of the European Parliament and of the Council of 9 March 2011 laying down harmonised conditions for the marketing of construction products and repealing Council Directive 89/106/EEC, *OJ* L88/4.4, 2011.
11. Trevisi, R.; Risica, S.; D'alessandro, M.; Paradiso, D.; Nuccetelli, C.: Natural radioactivity in building materials in the European Union: a database and an estimate of radiological significance, *J. Environ. Radioact.*, 2012, **105**, 11–20.

GAMMA-RAY SPECTROMETRY MEASUREMENT OF CONCRETE SAMPLES FROM NUCLEAR DECOMMISSIONING FACILITIES - VALIDATION OF THE LABSOCSTM EFFICIENCY CALIBRATION AT LOW ENERGIES

N. Dehbi*, C. Véronneau, JC. Piroux, F. El Yadari, G. Gasse and P. Schenckbecher

Laboratoire de Caractérisation et d'Analyses LCA d'ASTERALIS (VEOLIA NUCLEAR SOLUTIONS), France

*nabil.dehbi@veolia.com

ABSTRACT

Proper characterization and segregation of materials arising during decommissioning operations are very important factors in waste minimization. Characterization helps develop a complete understanding of the radiological characteristics of these materials in order to enable them to be segregated and to be sent for selected disposal. A number of techniques for measuring the radioactivity of materials generated in decommissioning operations are available. Failure in control procedures at any step can cause important consequences, not only in follow-up steps, but, in some cases, may result in generating waste packages which are not compliant with the waste acceptance criteria for long term storage or disposal. In this work we propose three different methodologies for the adjustment of the samples attenuation coefficients in order to validate the efficiency calibration curves generated by Mirion LabSOCS™ (Laboratory Sourceless Calibration Software). The three methodologies can be applied in the case where it is required to perform **γ**-rays spectrometry measurements at energies below 200 keV, on concrete samples of unknown chemical composition.

1 INTRODUCTION

γ-ray spectrometry using high-purity germanium (HPGe) detectors is a widely used technique in determining the radioactivity of various radionuclides in the field of decommissioning of nuclear facilities. Being a non-destructive technique, this method possesses advantages in multi-element analysis, simplified sample preparation without radiochemical separation process, and applicability for precise quantitative determination of γ-ray emitting radionuclides content in samples.

When measuring samples using HPGe detector, calibration of the system is a complicated process, because the counting efficiency depends on both the geometry of the sample and the attenuation coefficient of the compound. The activity of samples in γ-ray spectrometry is derived from the efficiency curve ε(E) determined for a calibration source (standard). The most accurate method to determine the activity of each radionuclide is to use an adequate standard source with very similar geometrical dimensions, density and chemical compositions to the sample under measurement. In practical applications, it is not always so as the sample and source materials might be entirely different. However, the preparation of special standards for each case is a time consuming process and that is why these specific standards are not available. It's why some authors use Monte Carlo code simulation through LabSOCS™ software to calculate photon transmittance and photo peak efficiency for energies above 200 keV, but, it doesn't work well for low energies because the accuracy of efficiency in this low region depends strongly on adequate knowledge of the elementary chemical composition of the source matrix in addition to the detailed characteristics of the germanium semi-conductor detector.

Environmental and Decommissioning samples may vastly differ in their chemical composition, their density ranging from near 0 to about 2,4 g cm^{-3}. That is why the self-attenuation correction is required to account for the differences in self-attenuation between the source and the standard.

In the case where it is required to perform, measurements at energies below 200 keV, on concrete samples of unknown composition, it is necessary to adjust the attenuation coefficient of the sample and to validate the calibration curve generated by LabSOCS™. In this work we propose different methodologies to do so.

2 INSTRUMENTS AND METHODS

The γ-ray spectrometry system used at the LCA ASTERALIS Laboratory is composed of a semiconductor detector consisting of germanium crystals, BEGe broadband energy with carbon window (Carbon Epoxy). The detector has been characterized by Mirion. It is a coaxial detector type (N) large area ($3800 mm^2$) and thickness 31,5mm for the simultaneous detection of low and high energy photons. The γ spectrometry system is composed also of a LYNX-MCA multichannel analyser with integrated digital electronics, with a processing time of 0,2 to 51 µs and a digital spectrum stabilizer; a Cryo-Cycle II system with electric hybrid cooling / liquid nitrogen with a range of more than 1 year or several days in case of power failure; a 10 cm lead-weight lead castle to protect against background noise due to ambient radiation, with sliding doors on the front. Inside the lead castle is a 2mm copper thickness to stop Lead X-rays;

The system uses Genie 2000 spectrometry software and algorithm that allows the acquisition, visualization and analysis of spectrometric data on a computer.

The LabSOCS™ modeling software is used for efficiency calibration without the use of radioactive sources. However, when it is required to measure solid samples with unknown chemical composition, especially in the case of the determination of low energies radionuclides, it is necessary to adjust and validate the LabSOCS™ model. This adjustment must take into account the self-attenuation by the sample matrix. For this, one of the following three methods is used:

2.1 Method 1

Determination of the self-absorption coefficient by transmission measurements using the radioactive source of ^{241}Am; Below 200 keV, the self-attenuation correction factor F_c is the product of two factors: F_{cd} factor related to the density of the sample and F_{cm} related to the atomic composition of the matrix. The validation of this F_{cm} factor for a sample against the calibration model by LabSOCS™ is performed by the experimental determination of the linear attenuation coefficient of the sample, which is compared to that of the calibration model. This experimental determination is carried out by measurements of transmission of a radioactive source through the sample (see Figure 1).

2.2 Method 2

Use of ^{137}Cs (if present in samples), as internal standard (cesium has lines at 661,7 keV and 32 keV); This method can be used for samples that contain ^{137}Cs in measurable quantities, in addition to the low-energy radionuclide that is being measured. Self-attenuation varies very little according to the chemical composition of the sample for energies higher than 200 keV. The 661,7 keV line can therefore be used to determine the ^{137}Cs content of the unknown sample. The aim is to measure the same activity at 32 keV.

2.3 Method 3

Measurement of the same sample on variable heights; This method consists of preparing two identical containers, into which two different amounts of unknown sample to be measured are introduced, so that the two filling heights are significantly different from each other. The two geometries are modelled by LabSOCS™, indicating the exact values of the densities of the samples, the masses introduced, the filling heights, and choosing a chemical composition model that seems adapted to the nature of the sample (for example, a sandy soil).

The two samples are then counted for a time sufficient to obtain a good counting statistics on the peaks of interest. The aim is to measure the same activity (in both cases) of a radionuclide emitting at energy below 200 keV.

Figure 1 *(Top) descriptive diagram of the method of transmissions. (Middle) device used for measuring the transmission of the collimated photon flux. (Bottom) Overview of the Gamma Spectrometry Measurement Chamber.*

3 RESULTS AND DISCUSSIONS

3.1 Effect on the matrix

To highlight the influence of the composition of the sample, we calculate the efficiency of the detector, by fixing a constant density, for different matrix: soils, sands, concretes, glass and polyethylene. The calculations were performed by the LabSOCS™ software. We have found, for the concrete and sand matrix, that there is more variation of the correction factor for low energies (<200 keV) than for high energies (> 200 keV). For energies below 200 keV, the efficiency of the measurement system strongly depends on the chemical composition of the samples. This is due to the fact that at these energies, the self-attenuation of radiation by the material constituting the samples is not negligible and varies greatly according to the energy of the photons and the atomic number of the elements present. In the case where it is desired to perform measurements at energies below 200 keV, on samples of unknown composition, it is necessary to adjust the attenuation coefficient of the sample and to validate the calibration curve generated by LabSOCS™, by one of the techniques discussed in this work.

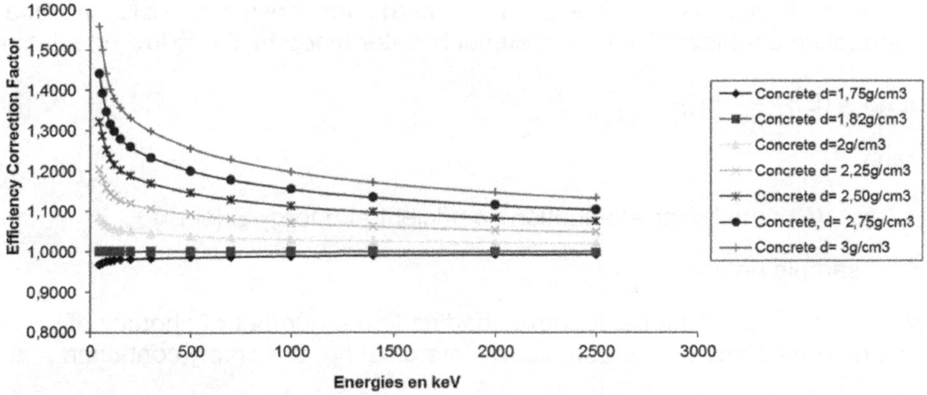

Figure 2 *Variation of efficiency versus energy for different concrete densities.*

3.1.1. Application of the method of determination of the self-attenuation coefficient by transmission measurements. In the case of measurement of solid samples (concrete, soil, sediment, sand) at low energies (<200 keV), attenuation coefficients are determined by the transmission measurement of a collimated flux of photons from ^{241}Am (59,54 keV) point source.

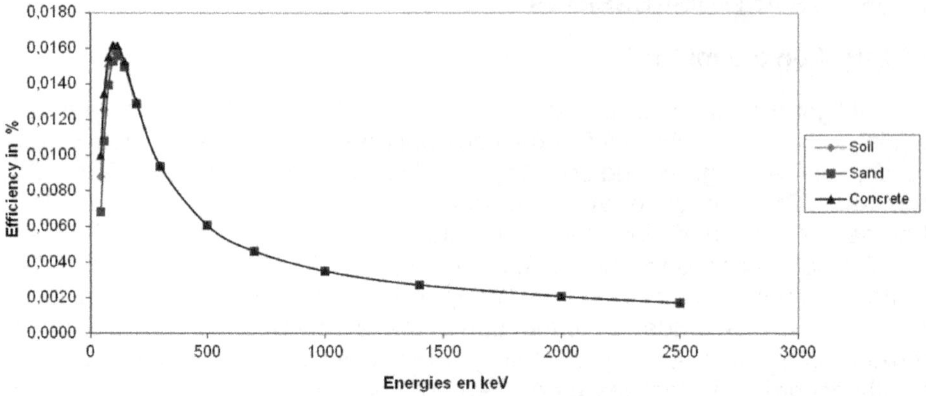

Figure 3 *Variation of efficiency versus energy for fixed density for different matrix (concrete, soil, sand).*

In order to confirm this method, we applied it to materials of known chemical composition: KCl, Na_2SO_4 and $MgSO_4$. The materials were packaged in flat boxes BTE60 (also known as petri dishes) in polystyrene with interlocking lid. From the transmission measurements made on these materials, the linear attenuation coefficient of each material is determined by the following relation:

$$\mu_{L\ sample}(E) = \frac{1}{X} \cdot \ln\frac{N_{v1}}{N_{p1}} \qquad (1)$$

With:

$\mu_{L\ sample}$ **(E)** : the linear attenuation coefficient at energy E (cm^{-1})

X : sample height.

N_{v1} : gross area of the peak corresponding to the number of photons of energy E detected by the detector by intercalating the empty container.

N_{p1} : gross area of the peak corresponding to the number of photons of energy E detected by the detector by intercalating the container full of sample.

In order to validate the LabSOCS™ model, it is necessary to compare the linear attenuation coefficients $\mu_{L\ sample}$ of the sample (E) to be analysed with that of the LabSOCS™ calibration model, $\mu_{Létalon}$ (E), for concrete or other material. The linear attenuation coefficients of the LabSOCS™ Soil Model (Standard) are calculated theoretically and determined from the chemical composition (chemical formula) of the LabSOCS™ standard material based on data from the XCOM mass attenuation coefficient tables: " Photon Cross Sections Database "of NIST (National Institute of Standards and Technology). The results of the measurements carried out according to the transmission method, with a source of ^{241}Am (geometry: flat cylindrical box) are presented in Table 1.

Table 1 Results of the application of the transmission method to three different materials

Material	Mass in g	Density in g cm^{-3}	Sample (X) in cm	N_{v1}	N_{p1}	μ_L sample cm^{-1}	μ_L standard cm^{-1}	Deviation (%)
KCl	70,22	1,17	1,49	23400	10200	0,557	0,589	-5,4
Na_2SO_4	80,81	1,34	1,49	23400	14600	0,315	0,336	-6,3
$MgSO_4$	97,94	1,63	1,49	23400	12700	0,409	0,425	-3,8

The linear attenuation coefficient of the concrete sample at energy (E), is calculated according to the standard ISO 18589-3.

An example of theoretical calculation of the linear attenuation coefficient of a LabSOCS™ Soil Calibration Model (density ρ = 1,53 g cm^{-3}) for the energy 59,54 keV is presented below (Table 2):

Table 2 Example of the calculation results of the linear attenuation coefficient of a LABSOCS™ soil calibration model

Elementary Composition of the soil LabSOCS™ Model	*Mass attenuation coefficient $\mu_m=\mu/\rho$ (cm^2/g)	Mass fraction w_i (%)	Linear attenuation coefficient μ_L of the standard soil LabSOCS™ Model (cm^{-1})
H	0,3260	0,17 %	0,386 cm^{-1}
C	0,1753	4,63 %	
O	0,1907	48 %	
Mg	0,2570	0,8 %	
Si	0,3207	38,36 %	
Ca	0,6578	8,05 %	

* *NIST data*

The coefficient of linear attenuation of the soil at energy (E), is calculated according to the standard ISO 18589-3 by the following equation:

$$\mu_L(E)=\left[\sum_i w_i \cdot \mu_{m,i}(E)\right] \cdot \rho \tag{2}$$

With: ρ the density of the soil or the material to be modelled. w_i is the mass fraction of each element i.

The results in Table 3 show that the soil model can be used for IARMA-001 soil, (difference of 4,7%) on the linear attenuation coefficient), but not for the IRSN 131SL300 sediment (difference of 31%), for which another non-standard model have to be used.

Table 3 Linear attenuation coefficient for two reference materials from IRSN and IARMA

CRM	Mass in g	Density in g cm^{-3}	Sample height (X) in cm	N_{v1}	N_{p1}	$\mu_{L\ sample}$ in cm^{-1}
IRSN 131SL	83,8	1,39	1,49	23400	15400	0,279
IARMA-001	88,9	1,48	1,49	23400	12800	0,405

3.1.2 Application of the method using ^{137}Cs as internal standard. The results of ^{137}Cs analysis in the IARMA-001 CRM at 661,7 keV and at 32 keV are presented in Table 4. The ^{137}Cs measurement results obtained from the two 661,7 keV and 32 keV lines are very close, that means that the calibration curve generated by LabSOCS™ is valid for this material with unknown chemical composition.

Table 4 Results of ^{137}Cs measurements in the IARMA-001 reference material at 661,7 keV and at 32 keV

Energies (keV)	661,7 keV	32 keV
Sample mass (g)	70,66	70,66
Measured activity (Bq/kg)	50 ± 6,6	52 ± 11
Reference value IARMA (Bq/kg)	52 ± 2,6	52 ± 2,6

3.1.3 Application of the variable heights method. The results of ^{137}Cs analysis in IARMA-001 soil CRM at different heights from the 32 keV line are presented in Table 5. The ^{137}Cs measurement results obtained from the 32 keV line with samples at two different filling heights (1,60 cm and 4,45 cm) in the SG50 flasks are very close: this is possible because LabSOCS™ correctly takes in to account the self-attenuation for this material that we don't know the chemical composition.

Table 5 Results of 137Cs measurements in the IARMA-001 reference materials at different heights at 32 keV

Energies (keV)	32 keV	32 keV
Sample mass (g)	26,35	70,66
Fill height of the sample in SG50 (cm)	1,60	4,45
Measured activity (Bq/kg)	51 ± 7	52 ± 11
Reference value IARMA (Bq/kg)	52 ± 2,6	52 ± 2,6

4 CONCLUSIONS

As a result of this work, the following conclusions are made in the case where it is required to perform measurements at low energies using LabSOCS™ efficiency calibration, on concrete samples of unknown chemical composition. In this work, we studied and validated three different methodologies to validate the LabSOCS™ standard model. The recommended method following this study is to adjust the attenuation coefficient of the LabSOCS™ standard model versus the real attenuation coefficient of the concrete sample determined by the transmission measurements (method 1). We also validated the two other methods: one by using ^{137}Cs as internal standard and another one by performing measurements for variable heights of the sample.

Those methodologies were validated by participating to IRSN Inter-laboratories Comparison Exercises and IARMA Proficiency Tests for the measurements of low energy γ-emitters radionuclides (Am-241, Cd-109, Pb-210, Th-234...) in solid samples.

The Characterization and Analysis Laboratory of ASTERALIS (Veolia Nuclear Solutions) is accredited on this method (according to the standard ISO/CEI 17025) by the French Accreditation Body COFRAC (Testing N° 1-5750: scope available on www.cofrac.fr) recognized by the International Laboratory Accreditation Cooperation (ILAC) , the International Accreditation Forum (IAF) and the European Cooperation for Accreditation (EA).

REFERENCES

1. A. E. M. Khater and Y. Y. Ebaid, A simplified gamma-ray self-attenuation correction in bulk, *Appl. Radiat. Isot.*, 2008, **66**, 407–413.
2. F. L. Bronson, Validation of the accuracy of the LabSOCS software for mathematical efficiency calibration, *J. Radioanal. Nucl. Chem.*, 2003, **255**, 137–141.
3. M, Długosz-Lisiecka and H. Bem, Fast procedure for self-absorption correction for low γ energy radionuclide ^{210}Pb determination in solid environmental samples, *J. Radioanal. Nucl. Chem.*, 2013, **298**, 495–499.

ELECTRON MICROSCOPIC ANALYSIS OF RADIOCAESIUM-BEARING MICROPARTICLES IN LICHENS COLLECTED WITHIN 3 KM OF THE FUKUSHIMA DAI-ICHI NUCLEAR POWER PLANT

Terumi Dohi*[a], Hisaya Tagomori[a], Yoshihito Ohmura[b], Kenso Fujiwara[a], Seiichi Kanaizuka[c] and Kazuki Iijima[a]

[a] 10-2, Fukasaku, Miharu, Tamura-gun, Fukushima 963-7700, Japan, Fukushima Radiation Measurement Group Fukushima Environmental Safety Center Japan Atomic Energy Agency. [b] 4-1-1 Amakubo, Tsukuba, Ibaraki 305-0005, Japan, Department of Botany, National Museum of Nature and Science. [c] 3129-29, Hirahara, Muramatsu, Tokai-mura, Ibaraki 319-1112, Japan, Nuclear Engineering Co., Ltd.

*Corresponding author: dohi.terumi@jaea.go.jp

ABSTRACT

In order to investigate statistically the physicochemical properties of the radiocaesium-bearing microparticles (CsMPs) emitted from the Fukushima Dai-ichi Nuclear Power Plant, an efficient method for collecting and separating the microparticles is required. In this study we established a separation method for CsMPs from lichens using electron microscopic analysis. Lichens were first digested within hydrogen peroxide and then an automatic particle analyser system based on FE-EPMA was used for detecting CsMPs within lots of mineral-like and metalic particles quickly. The elemental composition and radiocaesium activities of the separated CsMPs were measured, and the results suggest that the digestion treatment did not influence the CsMPs characteristics. An inhomogeneous elemental distribution was found in some CsMPs by two-dimensional elemental analysis using FE-EMPA. Our methods for separation of CsMPs and analysing their elemental composition are expected to be useful for grasping the chemical composition trends of CsMPs statistically as first step before more detailed analyses such as TEM analysis.

1 INTRODUCTION

Mainly volatile radionuclides were released into the atmosphere during the Fukushima Dai-ichi Nuclear Power Plant (FDNPP) accident in March 2011.[1] Local radioactivity depositions around the FDNPP and radioactive plume traces were formed due to the multiple releases.[2] In order to clarify the deposition characteristics of the different plumes, and the subsequent behaviour and migration of radiocaesium in the environment, clarification is needed for the physicochemical states of the radiocaesium.

In recent years, microscopic particles containing radiocaesium, so called "radiocaesium-bearing micro particles" (CsMPs)[3,4] have been reported as one of the physicochemical states.[5-8] The physicochemical properties of CsMPs will influence their migration in the environment, and also retain information on the processes that occurred during the reactor accidents. Thus, it is necessary to investigate their chemical characteristics statistically, e.g. elemental composition, internal structure, solubility etc. However, gathering such information is difficult because (1) radiocaesium is thought to have mainly been emitted in the gaseous form, and transported and deposited as sulphate aerosols,[9] (2) many processes are needed to collect and separate CsMPs from soils and other environmental samples, e.g. repetition of autoradiography, differentiation of CsMPs consisting mainly of Si and O from an enormous number of aluminosilicate and/or metal minerals, and subsequent separation of the CsMPs. Previous studies have identified CsMPs by electron microscopy and measured their radiocaesium activity with gamma spectroscopy.[3-8,10-12] It takes a lot of time and effort to search for and distinguish CsMPs from other mineral particles using an electron microscope. An efficient method to collect and separate CsMPs from environmental samples is required.

Here we focused on lichens as they are known to retain radionuclides for a long period and capture micro particles from the air.[13,14] By sampling lichens at a height greater than 1 m above the ground, terrigenous contamination by resuspended soil particles could be avoided.[15] This means the content of aluminosilicate particles in lichen samples is expected to be lower than in soil, litter and herbaceous vegetations at the ground surface. In addition, the propensity of lichens to retain particles prevents the loss of CsMPs when collecting, moving and downsizing samples. Past studies have inferred the presence of radioactive particles in lichens because autoradiographs showed hot spots.[16,17] In our previous study we identified CsMPs on lichen thalli using a field emission electron probe micro analyser (FE-EPMA) with mapping analysis.[18] However, some CsMPs in lichens may be difficult to find if they are hidden by an uneven lichen surface or hidden inside thalli.

The purpose of this study was to establish an efficient separation method for CsMPs captured by lichen using electron microscopic analysis and to investigate the characteristics of those CsMPs.

2 MATERIALS AND METHODS

2.1.1 Sampling points. Lichen samples were collected at four sites within 3 km of the FDNPP, located within plumes to the west (W-N, W-N2) and the south directions (S-W, S-E1), between Jan 2016 and Jan 2017 (Figure1). Samples of *Parmotrema tinctorum* (Nyl.) Hale, *P. clavuliferum* (Räsänen) Streimann and *Punctelia borreri* (Sm.) Krog, common species growing on trees within the area, were collected at heights of 1 m or greater. The species were identified based on the method in T. Dohi *et al.*, 2015.[19]

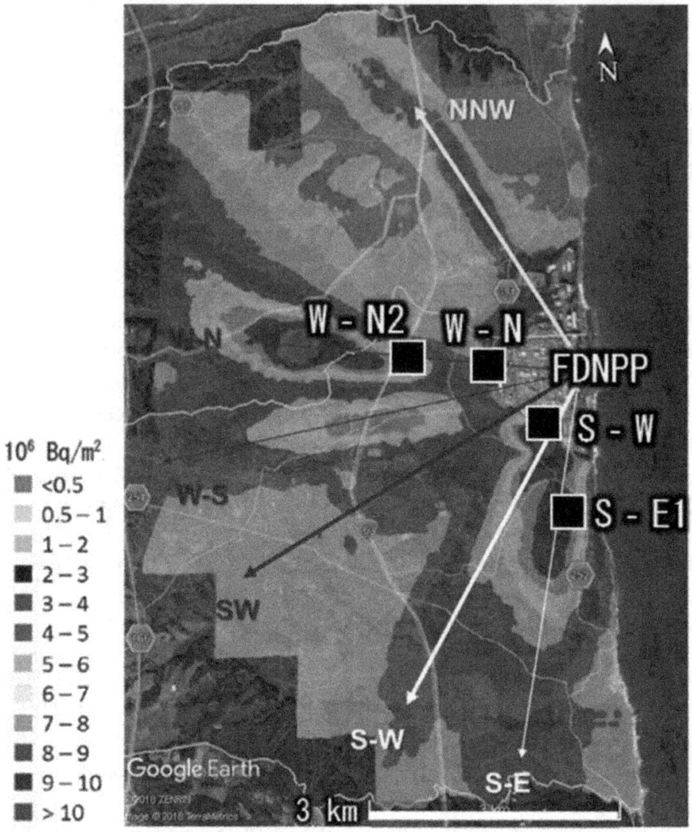

Figure 1 *Sampling points. Contour map showing 137Cs deposition density level measured by unmanned helicopter. This level was decay corrected to 31 July 2013. (Airborne monitoring data in Database for Radioactive Substance Monitoring Data (https://emdb.jaea.go.jp/emdb/portals/b1020202/). Arrows indicate plume traces.*

2.1.2 Digestion treatment and preparation for electron microscopic analysis. Each of the process described below is outlined in Figure 2. First the lichen samples were exposed for 15 minutes to an imaging plate (IP) (BAS-SR2040, FUJI FILM) and the autoradiograph images were scanned by a laser scan analyser (Typhoon FLA7000, GE Healthcare) (Figure 2.1).

Each lichen sample was cut down using a razor to a circa 10 mm by 10 mm sized fragment (about 0.01 g, dry weight) which included a hot spot visible on the autoradiograph image (Figure 2.2) and then digested using 5 ml of 30% hydrogen peroxide in a 15 ml centrifuge tube for 1 hour at room temperature (Figure 2.3). The tube was then kept at 100°C for 2 hours. After 30 minutes cooling, the solution was centrifuged at 1300g (3000 rpm) for 10 minutes (CR-22, Hitachi Koki) and the supernatant was discharged. The residual sample was washed by addition of 5 ml ultrapure water followed by centrifugation three times. Finally, 2 ml ultrapure water was added to the tube (Figure 2.4). The sample was lightly stirred and the whole suspension was dropped onto a glass slide by micropipette. The suspension on the glass slide was dried at 60°C using a hot plate for 1 hour (Figure 2.5).

The glass slide with precipitates was wrapped in parafilm and exposed to the IP for 10 minutes to identify the position of the CsMP. The part of the parafilm around the spot was cut out with a razor. Next carbon tape was pasted to the cutting area to take away the precipitates containing the CsMP, twice at the same position, and transferred to another glass slide and cut down to around a 2 mm x 2 mm square fragment under a two-step process with autoradiography (Figure 2.6). Finally, the fragment of carbon tape with the CsMP was moved to an aluminium sample table (12.5 ϕ × 10 mm height, JEOL) for electron microscope analysis. The sample was carbon-coated to 30 nm thickness by carbon deposition equipment (IB-Z17052THL, JEOL).[20]

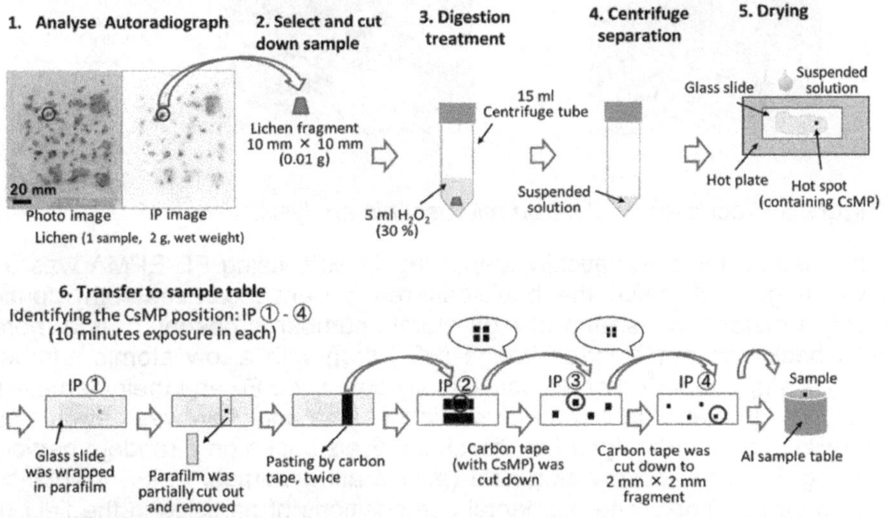

Figure 2 *Processes of digestion and preparation for electron microscope analysis.*

2.1.3 Electron microscopic analysis. The sample table was set in a holder for position coordinate sharing between a field emission electron probe micro analyser (FE-EPMA) (JXA-8530F, JEOL), a focused ion beam (FIB) (JIB-4000, JEOL) and a micro manipulator (Axis Pro FC, Micro Support Co., Ltd & JEOL) (Figure 3.1). The position coordinate share holder was used with some control software (IB-67030SLS, JEOL) common to all the equipment. This allowed the sharing of coordinate information of candidate CsMPs between the equipment.

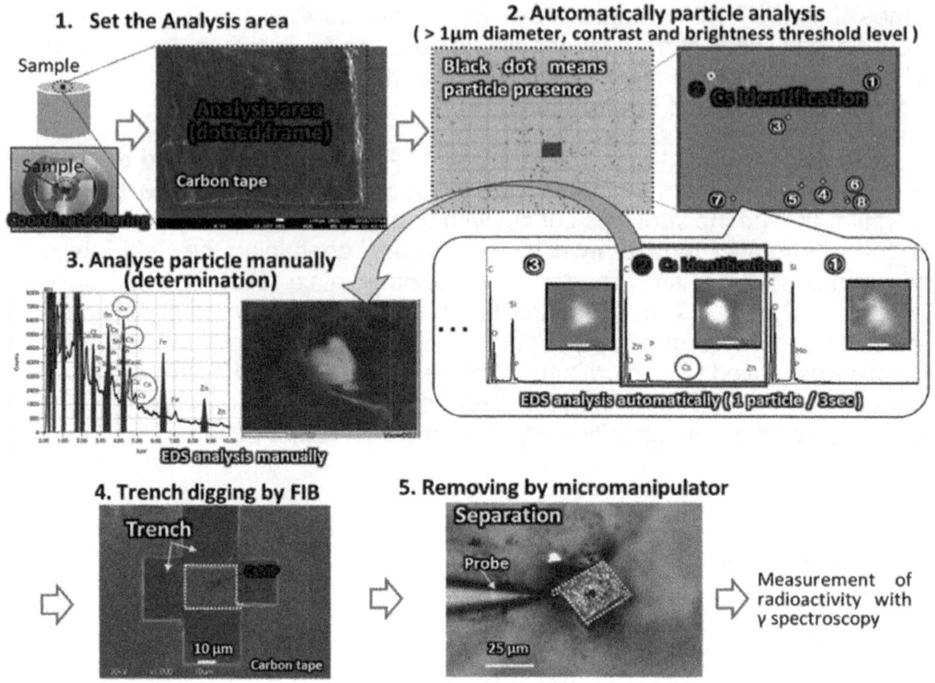

Figure 3 *Processes of electron microscopic analysis.*

The method for automatically identifying CsMPs using FE-EPMA was as follows. In general, since the backscattered current increases with atomic number, a material consisting of high atomic number is observed as a more brightly backscattered electron image (BEI) than with a low atomic number material. CsMPs reported previously usually contained Fe and their diameters were mostly greater than 1 µm.[21] A particle search condition was determined using BEI contrast and brightness threshold levels based on a random particle containing Fe in the observation field (accelerating voltage 15 kV, irradiation current around 50 nA). The elemental compositions of particles in the field of view were measured using energy dispersive X-ray spectrometry (EDS) analysis in FE-EPMA. Particles composed of atoms with atomic number equal or greater than that for Fe, and with size greater than 1 µm, were searched for by the control software (EX-35110, JEOL) (conditions: magnification 500, number of pixels: 512×384, live time of EDS: 3 seconds). A particle expected to contain Cs was selected automatically based on the results of the search (Figure 3.2). The candidate particle was then analysed manually with EDS to see if the elemental composition matched that for a CsMP (Figure 3.3).

The elemental composition of the whole CsMP surface was analysed by wavelength dispersive X-ray spectrometer (WDS). The two-dimensional elemental distribution of the CsMP was measured using the WDS detector. Mapping analysis was carried out with the WDS detector (condition: number of pixels 266×192, measuring time 10 milliseconds per point).

After the FE-EPMA analysis, the sample was transferred to the FIB. In the FIB a trench was dug to create a circa 20 μm² square area around the CsMP using a Ga ion beam (accelerating voltage 30 kV, irradiation current 30 nA) (Figure 3.4). The fragment of carbon tape containing the CsMP was removed using a tungsten probe (5 μm ϕ, Micro Support Co. Ltd) micro manipulator assisted by the coordinate sharing software (Figure 3.5).

2.1.4 Measurement of radioactivity with γ spectroscopy. The radioactivity of Cs in the CsMPs was determined by the intensity of the gamma-ray peaks at 604 keV for ^{134}Cs and 662 keV for ^{137}Cs when measured with a high-purity Ge (HPGe) semiconductor detector (GMX40P4-76, Seiko EG&G ORTEC and GC2518, CANBERRA, relative efficiencies of 40% and 25% respectively). Efficiency calibration was performed using standard point sources of ^{134}Cs (CZ402, Japan Radioisotope Association) and ^{137}Cs (CS402, Japan Radioisotope Association). These sources were installed at the bottom of a V1 container.[22] CsMP samples were presented to the HPGe detector with the same geometry as that for the standard sources. The measuring time was set to ensure over 500 net counts from ^{134}Cs.

3 RESULTS AND DISCUSSION

3.1.1 CsMPs separation by combined lichen digestion treatment and automatic particle analysis with FE-EPMA. About 100 to 2000 insoluble particles were obtained per circa 10 mm by 10 mm sized fragment of each lichen sample after hydrogen peroxide digestion treatment and automatic particle analysis using FE-EPMA. Their sizes ranged from 1 to 50 μm diameter. In the case of 2000 insoluble particles, one CsMP was differentiated after approximately 4.5 hours of automatic searching. After this initial detection, it took several more hours to separate it from the other particles on the carbon tape with the FIB and the micro manipulator. In total it took up to 3 days from autoradiography to CsMP separation (Figures 2 and 3). Comparing our automatic method with a purely manual separation method, the latter took several days (around 100 particles analysed per day) to search for the CsMP. However even if CsMPs were indicated on the autoradiographs of the lichens, it may not be possible to find them manually. This may be because they were hidden inside tissues or within lichen thalli. Because separation of CsMPs from environmental samples depends on the radioactivity and size of the particle, and the type of sample material, it is difficult to specify and compare the times required for CsMP separation from different samples. However, in the case of soil samples taken near to the FDNPP, it typically takes two weeks or more to separate one CsMP.[10-12]

Other organic materials, e.g. moss, have the ability to capture airborne particles like lichen.[23] Our combined digestion and automatic analysis method could be used for moss, despite moss having a three-dimensional structure. On the other hand, in the case of higher plants containing lots of silica, so-called "biogenic silica" like poaceous grasses and other plants,[24,25] and litter containing lots of soil minerals, additional treatment steps may be required to implement our method, such as reducing the amount of sample as the number of insoluble particles increases. The method of automatic particle identification is also expected to be applicable for other particles, e.g. PM2.5, PM10, heavy metals and other contaminants.[26]

3.1.2 Electron microscopic analysis of CsMPs. We obtained 12 CsMPs from lichens collected in the west and south plume directions by applying the above method. Figure 4 and Table 1 show that their diameters were around 2 μm (range 1.4 – 4.4 μm) and their shapes were both spherical and non-spherical. The elemental compositions at the surface of the CsMPs identified by FE-EPMA WDS / EDS analysis were Si, O, Zn and Fe coexisting with Cs (Table 1). These CsMPs, mainly containing Si and O, were similar to the CsMP found on a lichen thallus without any treatment.[18] That lichen was also *P. tinctorum* collected 4 km south-west of the FDNPP (specimen No. TD132).[19] Furthermore, the characteristics of the CsMPs in this study were similar to other CsMPs separated from soil and gravel samples from within 4 km of the FDNPP without chemical treatment.[10-12] These studies reported CsMPs with diameters ranging from 2.0 to 17.3 μm and containing Si, O, Fe, Zn and Cs as major elements. According to the classification system outlined by Satou *et al.*, 2018,[27] the CsMPs separated from our lichen samples are broadly "type A" particles, based on their size and constituent elements. Internal structure analysis using transmission electron microscopy (TEM) indicated that these type A particles are basically silicate glass.[4,6,10-12] It has been suggested that type A particles are largely insoluble to water.[5] The CsMPs we obtained from lichen after H_2O_2 treatment all had (i) sizes (mostly 1–10 μm), (ii) shapes (spherical and non-spherical) and (iii) elemental compositions (mainly Si, O, Zn and Cs) that are similar to the type A particles that were found by others.[4-6,10-12,27-28] Therefore, the hydrogen peroxide treatment is not considered to have changed the CsMPs characteristics in a significant way. In future we plan to check this conclusion by subjecting a CsMP obtained by conventional separation to H_2O_2 digestion, and comparing its properties before and after the treatment. Either way, our results demonstrate that lichens captured CsMPs, like soil and other environmental samples.

Table 1 *Detected elements within CsMPs.*

Sample Number	Sample name	Particle shape	Diameter (with maximum length) (μm)	Elements (Detected by WDS and/or EDS) Except for C
1	W-N_L1_1	non-spherical	2.1	Ba, Sn, Mn, Ir
2	W-N_L1_3	non-spherical	2.3	Ba, Sn, Mn, Rb, P, In
3	W-N_L2_1	spherical	2.5	Sn, Mn, Cr, U, P
4	W-N2_3_L1_1_1	spherical	1.8	Ba, Sn, Mn, U, P, S
5	W-N2_3_L1_1_2	spherical	1.4	Ba, Sn, P
6	W-N2_3_L1_2_1	non-spherical	1.9	Mg, Al, Ca,Ti, Mn, Cr
7	W-N2_3_L1_2_2	non-spherical	2.0	Sn, P
8	W-N2_2_L2_2	non-spherical	2.7	Cs, I (* Only 2 elements detected)
9	S-W_1_L1_1	spherical	2.8	Ba, Sn, Mg, Mn, P, Al, Ca, Cr, Cu, Rb
10	S-W_1_L1_4_1	spherical	4.4	Mg, Al, P, Ca, Mn
11	S-W_1_L1_4_2	spherical	1.8	Ba, Sn, Mn, Al, P, Ca , P, In
12	S-E_1_L4	spherical	3.6	Mg, Al, Ca, Mn, Sn, In
Common elements of samples in this study (except for No. 8)				Cs, Zn, Fe, Si, O, K, Cl, Na
References (Detected by TEM and/or SR-XRF analysis) Except for NNW direction area form the FDNPP)				Ba, U, Mn, Rb, Mo, In, Sb, Te, Ag, Zr, Pb

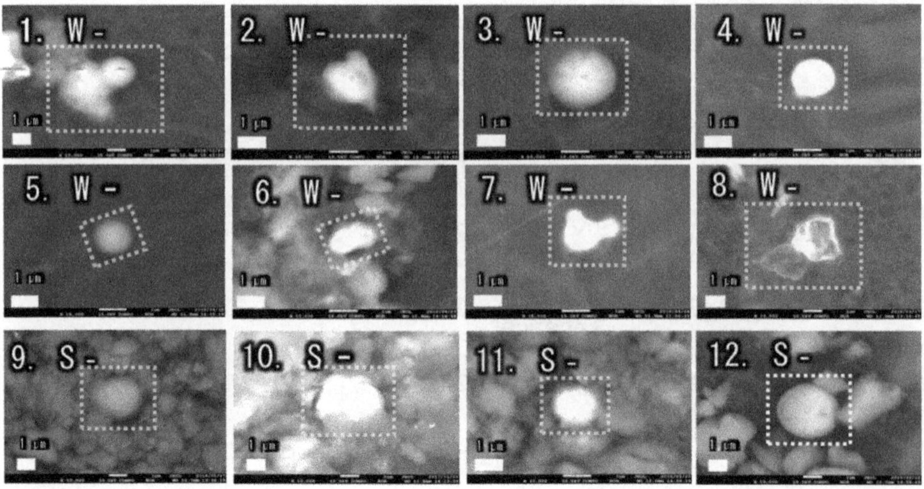

Figure 4 *Electron microscope images of CsMPs within dried suspension from lichens. Each CsMP is shown with a yellow outline. These images show back scattered images. Particles 1 to 8 were in lichens collected in the direction west of the FDNPP, and particles 9 to 12 were collected to the south. Particle 3 was found manually after the automatic search was stopped midway for readjustment of a search condition.*

Figures 5 and 6 show two-dimensional elemental distributions for two CsMPs from the 12 samples. Inhomogeneous elemental distributions were found for U, Sn, Fe and Cr (Figures 5 (d) and 6 (f) to (h)). This result, obtained by FE-EPMA, is consistent with results from other analysis methods. Kogure *et al.*, 2016 reported that Cs and Sn were distributed heterogeneously in CsMPs, and SnO_2 nanocrystals were observed at the surface based on TEM.[4]

Yamaguchi *et al.*, 2018 found nano particles containing Cr and Fe distributed inside a CsMP by internal structure analysis with TEM.[29] Trace elements reported were also detected such as Ba, Mn and Rb (Figure 5 (c) and (e) (Figure 6 (c) to (e), Table 1). These elements were detected only by trace element analysis, TEM and Synchrotron Radiation X-ray Analysis (SR-XRF).[3,4,6,8,10-12,21,27,29] Figures 5 shows the distribution of trace elements in our samples. These images are useful for establishing the relationships between the distributions of different elements. Cs and Ba have similar distributions, suggesting the possible presence of Ba as daughter nuclides of Cs.

The heterogeneous elemental distributions and trace elements found in the present study may imply that Fe-Cr minerals and SnO_2 crystals are incorporated within the CsMPs we analysed. The FE-EPMA mapping data (Figures 5 and 6) provide information pertinent for future studies seeking to clarify the surface / internal structure of CsMPs. In addition, the data are expected to helpful for grasping the main chemical composition and distribution of elements in our CsMPs as first step before detailed trace elemental analysis, e.g. using TEM.

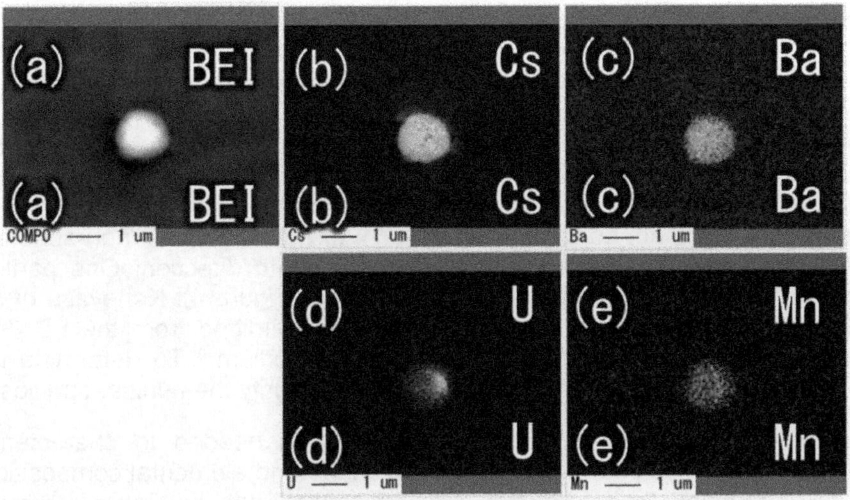

Figure 5 *Elemental distribution in CsMP collected at W-N2 (Figure 4.4). Image (a) shows a backscattered electron image (BEI) of the CsMP, and (b) to (e) show the elemental distribution. Image (d) indicates the local concentration of U.*

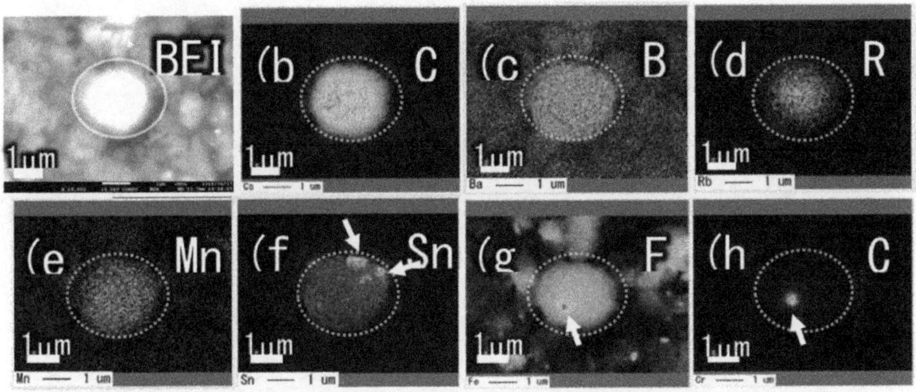

Figure 6 *Elemental distribution in CsMP collected at S-W (Figure 4.9). Image (a) shows a backscattered electron image (BEI) of the CsMP, and (b) to (h) show the elemental distribution. Image (f), (g) and (h) indicate the local concentration of Sn, Fe and Cr, respectively.*

3.1.3 Radioactivities of CsMPs. Five of the twelve CsMPs were separated and determined to have ^{137}Cs radioactivity in the range 0.214 ±0.005 to 6.660 ± 0.017 Bq on the measurement dates (Jan. to Jun. 2018). This corresponds to a range 0.254 ± 0.006 to 7.825 ± 0.020 Bq when decay corrected to 11 March 2011. The radioactivity ratios of $^{134}Cs/^{137}Cs$ on 11 March 2011 were 0.970 ± 0.062 to 1.142 ± 0.157 (Figure 7). For the other two CsMPs, only ^{134}Cs activity was checked. The identification of ^{134}Cs in the CsMPs shows that they are from the FDNPP accident. The $^{134}Cs/^{137}Cs$ ratios indicate they are derived from Unit

2 and/or Unit 3 of the FDNPP based on Nishihara *et al.*, 2012[30] and Chino *et al.*, 2016.[2] Figure 7 shows that samples No.1 and No. 12 (Table 1 and Figure 4) may be derived from Unit 3 and Unit 2, respectively. These CsMPs had different shapes (Figure 4). Spherical type particles suggested that the melted materials solidified rapidly in the air or in another situation.[21] Thus, it is possible that samples No.1 and No.12 may have different mechanisms or environmental factors associated with their formation.[21] On the other hand, their elemental compositions were not so different (Table 1). In addition, more non-spherical CsMPs were found in west direction than the south direction. One particle composed of Cs, I and C was found (Table 1 and Figure 4). Kaneyasu *et al.*, 2012 reported that CsI particles may have been emitted from the FDNPP, created by ion-induced nucleation of transport medium.[9] To determine the origin of our CsI particle, it is necessary to further clarify the nuclear species.

Collection and separation of many CsMPs is needed to characterize statistically the particle tendencies such as shape and elemental composition, and to understand their formation mechanisms. Future studies will require numerous CsMPs to clarify their physicochemical properties.[31] Our method of CsMP separation is a candidate for obtaining large numbers of CsMPs efficiently.

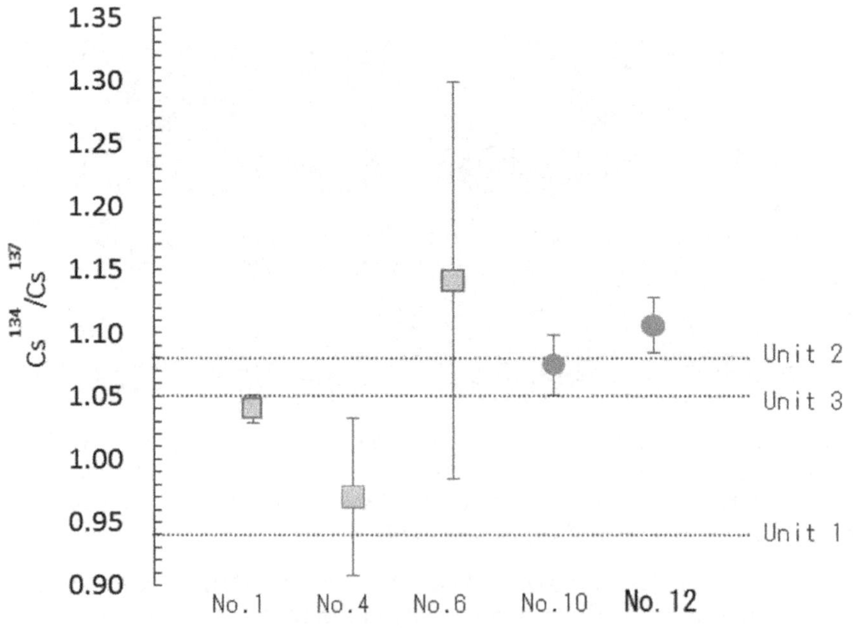

Figure 7 *134Cs/137Cs radioactivity ratio of five CsMPs separated from lichen samples. The activities were decay-corrected on 11 March, 2011. The 134Cs/137Cs radioactivity ratios for FDNPP units 1 to 3 were based on Ref. 30, 2 and 28.*

4 CONCLUSIONS

We showed for the first time that CsMPs can be separated from lichen with digestion treatment. FE-EPMA with automatic particle analyser system is useful for quickly detecting CsMPs greater than 1 μm diameter and radioactivity greater than 0.24 Bq radiocaesium (^{134}Cs + ^{137}Cs) from huge amounts of mineral like and metal particles. Thus, this method of microscopic analysis for obtaining CsMPs from lichen or other environmental samples is expected to be useful for characterizing the chemical composition of CsMPs statistically in future.

CONFLICTS OF INTEREST

There are no conflicts to declare.

ACKNOWLEDGEMENTS

The authors would like to express our gratitude to the local governments of Okuma Town and Tokyo Electric Power Company Holdings for permission undertake to our field investigations. We are grateful to Dr Alex Malins for proofreading our manuscript. We also thank Dr H. Yoshikawa and Mr Nakama for their support for our investigation. The authors wish to acknowledge Prof. Kogure for his advice on FIB processing. We appreciate Dr Utsunomiya, Dr Higaki, Dr Adachi, Dr Yamaguchi, Dr Okumura and Dr Satou for sharing information from their research with us.

This work was supported by MEXT, The Center of World Intelligence Project for Nuclear S&T and Human Resource Development, Number 273505. This work was also supported by MEXT/JSPS Number JP16K12627.

NOTES AND REFERENCES

1. G. Katata, M. Chino, T. Kobayashi, H. Terada, M. Ota, H. Nagai, M. Kajino, R. Draxler, M. C. Hort, A. Malo, T. Torii and Y. Sanada, *Atmos. Chem. Phys*, 2015, **15**, 1029–1070.
2. M. Chino, H. Terada, H. Nagai, G. Katata, S. Mikami, T. Torii, K. Saito and Y. Nishizawa, *Sci. Rep.*, 2016, **6**, 31376.
3. Y. Abe, Y. Iizawa, Y. Terada, K. Adachi, Y. Igarashi and I. Nakai, *Anal. Chem*. 2014, **86**, 8521–8525.
4. T. Kogure, N. Yamaguchi, H. Segawa, H. Mukai, S. Motai, K. Akiyama-Hasegawa, M. Mitome, T. Hara and T. Yaita, *Microscopy*, 2016, 451–459.
5. K. Adachi, M. Kajino, Y. Zaizen and Y. Igarashi, *Sci. Rep.*, 2013, **3**, 2554.
6. N. Yamaguchi, M. Mitome, K. Akiyama-Hasegawa, M. Asano, K. Adachi and T. Kogure, *Sci. Rep.*, 2016, **6**, 20548.
7. Y. Satou, K. Sueki, K. Sasa, K. Adachi and Y. Igarashi, *Anthropocene*, 2016, **14**, 71–76.
8. T. Ono, Y. Iizawa, Y. Abe, I. Nakai, Y. Terada, Y. Satou, K. Sueki, K. Adachi and Y. Igarashi, *BUNSEKI KAGAKU*, 2017, **66**(4), 251–261. In Japanese.
9. N. Kaneyasu, H. Ohashi, F. Suzuki, T. Okuda and F. Ikemori, *Environ. Sci. Technol.*, 2012, **46**, 5720–5726.

10. G. Furuki, J. Imoto, A. Ochiai, S. Yamasaki, K. Nanba, T. Ohnuki, B. Grambow, R. C. Ewing and S. Utsunomiya, *Sci. Rep.*, 2017, **7**, 42731.
11. J. Imoto, A. Ochiai, G. Furuki, M. Suetake, R. Ikehara, K. Horie, M. Takehara, S. Yamasaki, K. Nanba, T. Ohnuki, G. T. W. Law, B. Grambow, R. C. Ewing and S. Utsunomiya, *Sci. Rep.*, 2017, **7**, 5409.
12. Ochiai, J. Imoto, M. Suetake, T. Komiya, G. Furuki, R. Ikehara, S. Yamasaki, G. T. W. Law, T. Ohnuki, B. Grambow, R. C. Ewing and S, Utsunomiya, *Environ. Sci. Technol.*, 2018, **52**, 2586–2594.
13. M. R. D. Seaward, In: *Monitoring with Lichens: Monitoring Lichens.* Kluwer Academic Publishers, Dordrecht, 2002, 85–96.
14. T. H. Nash III, in *Lichen Biology*, 2008, 236–253.
15. R. Bargagli and P. L. Nimis, In: *Monitoring with Lichens: Monitoring Lichens.* Kluwer Academic Publishers, Dordrecht, 2002, 295–299.
16. T. Sawidis, K. Tsigaridas and L. Tsikritzis, *Ecotoxicology and Environmental Safety*, 2010, **73**, 1789–1796.
17. T. Dohi, K. Fujiwara, Y. Ohmura and K. Iijima, Abstract of *annual meeting of Atomic Energy Society of Japan (AESJ).* 2015, 566. In Japanese.
18. T. Dohi, K. Fujiwara, Y. Ohmura and K. Iijima, In: *Annual meeting of Atomic Energy Society of Japan (AESJ).* 2015
19. T. Dohi, Y. Ohmura, H. Kashiwadani, K. Fujiwara, Y. Sakamoto and K. Iijima, *J. Environ. Radioact.* 2015, **146**, 125–133.
20. H. Tagomori, Y. Ishii, S. Kanaizuka, T. Dohi and K. Iijima Abstract of *The Japan Society of Nuclear and Radiochemical Sciences 2017 Meeting*, 152.
21. K. Adachi, *Earozoru Kenkyu*, 2017, **32**(4), 255–2260. In Japanese.
22. Ministry of Education, culture, sports, science and technology Japan (MEXT), In: *Radioactivity measurement series*, 1992, **7**, 102–104. In Japanese.
23. P. Adamo, S. Giordano, D. Naimo and R. Bargagli, *Atmospheric Environment.* 2008, **42**, 346–357.
24. H. Motomura, N. Mita and M. Suzuki, *Annals of Botany.* 2002, **90**, 149–152.
25. H. A. Currie and C. C. Perry, *Annals of Botany*, 2007, **100**, 1383–1389.
26. T. Dohi, K. Fujiwara, Y. Ohmura and K. Iijima, Abstract of *annual meeting of Atomic Energy Society of Japan (AESJ).* 2018, 2E11. In Japanese.
27. Y. Satou, K. Sueki, K. Sasa, H. Yoshikawa, S. Nakama, H. Minowa, Y. Abe, I. Nakai, T. Ono, K. Adachi and Y. Igarashi, 2018, *Geochimical Journal*, **52**, 1–7.
28. S. Higaki, Y. Kurihara, H. Yoshida, Y. Takahashi and N. Shinohara, J. Environ. Radioact. 2017, 177, 65–70.
29. N. Yamaguchi, T. Kogure, H. Mukai, K. Akiyama-Hasegawa, M. Mitome, T. Hara and H. Fujiwara, 2018, *Geochemical Journal*, **52**, 1–14.
30. K. Nishihara, H. Iwamoto and K. Suyama, *JAEA-Data/Code*, 2012. In Japanese.
31. T. Okumura, N. Yamaguchi, T. Dohi, K. Iijima and T. Kogure, *Sci. Rep.*, 2018, **8**, 9707.

AUTOMATED PRODUCTION OF [^{18}F] PSMA-1007 USING 7.5 MEV CYCLOTRON: A FEASIBLE ALTERNATIVE FOR REMOTE SITES

Andreas Fesas,*[a] M. Reza Pourkhessalian [a] and Alexis Vrachimis [a,b]

[a]Department of Nuclear Medicine, German Oncology Centre, 4108 Limassol, Cyprus. [b]European University Cyprus, 2404 Nicosia, Cyprus.

* Andreas.Fesas@goc.com.cy

ABSTRACT

The necessity of targeting the prostate-specific membrane antigen (PSMA) with a radiotracer other than the generator-based [^{68}Ga] PSMA compounds, led to the production of the ^{18}F labelled PSMA-1007. The production of [^{18}F]PSMA-1007 was achieved using a 7.5 MeV cyclotron and the NEPTIS mosaic-RS automated synthesis module. After the first 36 batches, the mean percentage yield (decay corrected) was calculated to be 57.0 ± 10.1 % and the average activity in the final product vial reached 2.1 ± 0.5 GBq, which can be translated to approximately 3 patients per cassette; using a single PET/CT scanner. The time needed for the synthesis and quality control was 65 minutes. In order to reach an initial activity high enough to produce more than one dose per cassette and therefore for it to become economically viable to produce the radiopharmaceutical, 95 to 100 minutes of bombardment were required at a requested 5.5 μA target current. PET images have shown the expected results according to the literature.

1 INTRODUCTION

Prostate cancer (PCa) is the most common cancer in men and it can be characterised as the second most common cancer-related cause of death.[1-3]

When detected in early stages the survival rate can reach up to 95%.[4] Therefore, its early diagnosis is considered to be of the utmost importance. A plethora of positron emitting radiotracers have been developed to target PCa.[5-9] One of them is the positron-emitting gallium-68 (^{68}Ga) labelled prostate-specific antigen (PSMA) 617 or 11, which has been proven to be highly sensitive in detecting metastatic PCa.[7,8] However, ^{68}Ga is a generator-based radionuclide leading to limited production capacities. Furthermore, its decay properties (half-life of 68 minutes) and its higher positron energy with respect to ^{18}F (1.90 vs 0.65 MeV), which negatively influence the spatial resolution can be considered drawbacks.[6] Additionally, the emerging demands on GMP radiopharmaceutical production have led to a generator shortage. As a result, a radioactive fluorinated compound that could successfully target PSMA was of great interest due to the large number of facilities with immediate access to high activities of GMP-conforming ^{18}F production. Currently, [^{18}F]PSMA-1007 is considered the most promising candidate.[10-12]

BG75 Biomarker Generator is a low energy (7.5 MeV) self-shielded, proton beam, positive ion cyclotron[13] intended to satisfy patients' needs of 2-deoxy-[^{18}F]fluoro-D-glucose (FDG) using the principle of "a dose on demand" radiopharmaceutical production. This can be translated to "a cyclotron run" or "30 minutes of bombardment" for a dose.[14,15] The successful combination of BG75 with a third-party synthesis module, will boost the capabilities of the remote sites as they will no longer ne restricted to FDG and can offer additional PET examinations such as [^{18}F]PSMA-1007.

The challenges of the synthesis of [^{18}F]PSMA-1007 using a low energy cyclotron were mainly the low maximum target volume of 0.28 mL been irradiated, which translates to low initial activity introduced to the module.

We report herein on the feasibility of production of [^{18}F]PSMA-1007 using a BG75 cyclotron and NEPTIS mosaic-RS automated system by reporting our experience from the first 36 productions, the quality control methods used and some PET examples from the examinations performed.

2 MATERIALS AND METHODS

All chemicals, reagents and solvents needed for synthesis of [^{18}F] PSMA-1007 were purchased from ABX GmbH (Radeberg, Germany). A pre-conditioned Sep-Pak® Light Waters Accell™ Plus QMA cartridge was used for $^{18}F^-/^{18}O$ water separation. For the purification of the final product a Chromafix PS-H$^+$ (L) cartridge and a Chromafix C18$_{ec}$ (Macherev-Nagel; Duren, Germany) were used, both in a non-preconditioned form. The synthesis cassette and the reference compound of [^{19}F]PSMA-1007 used for the quality control were also purchased from ABX GmbH. HPLC solvents (HPLC grade) were purchased from Sigma-Aldrich. ^{18}F was produced by irradiation of ^{18}O enriched water (97%, ABX, GmbH; Radeberg, Germany) with a 7.5-MeV cyclotron according to $^{18}O(p,n)^{18}F$ nuclear reaction. The cyclotron's target current was maintained at 5.5 μA through the whole bombardment period. Irradiations were performed in the facility with the BG75 (ABT Molecular Imaging; Knoxville, TN, USA) 7.5 MeV proton beam, positive-ion cyclotron.

2.1 Radiosynthesis and quality control of [^{18}F] PSMA-1007

The radiosyntheses were performed using the NEPTIS mosaic-RS automated system (Ora; Philippeville, Belgium). The setup is illustrated in Figure 1. The positions of the cassette were loaded as follow: QMA along with the 10 mL conical column reservoir were placed on position 2. On position 3, a vial with 8 mL ethanol was placed and on position 5, a vial containing a solution of 1.6 mg of precursor in 2 mL DMSO was inserted. A 100 mL water bag containing 5.5 mL ethanol was connected at position 7. For elution of the cartridges, a vial containing 8.5 mL of 30% aqueous ethanol was mounted at position 8 and on position 9, a vial with 15mL of Phosphate buffer was placed. Position 10 was not used, and it was closed with a stopper. Two 30 mL syringes were attached on positions 4 and 14 and the inlet of the glass reaction vessel was connected at position 6. Its outlet at position 15. Detailed radiolabelling process was published by Cardinale *et al.*[11,16]

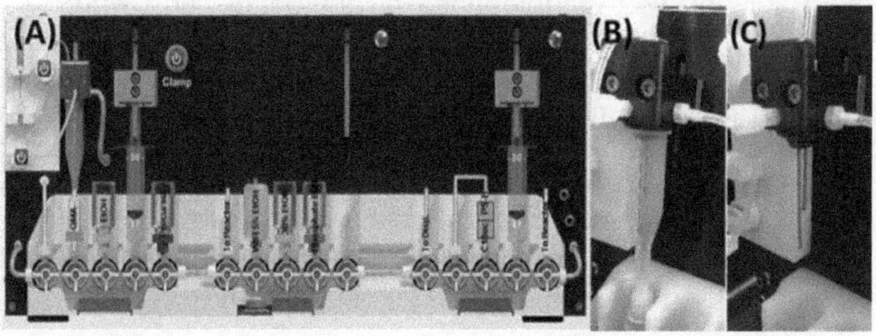

Figure 1 *(A) General setup of the radiosynthesis module for production of [^{18}F]PSMA-1007 Neptis mosaic-RS system. (B)Modification made to the delivery line. Two dotted lines show the usual delivery line size (right) and the modified delivery line (left). The lines can be seen clearly at (C).*

Modifications were made on the system to increase the production efficiency for a low initial volume of $^{18}F/^{18}O$ H_2O. The size of the line introducing the target water to the 10 mL conical column reservoir has been lengthened to bring it closer to the QMA cartridge, avoiding spreading income droplets to the sides of the reservoir, which could lead to a lower of target water passing through the QMA cartridge. Total initial volume of the target water that was passed through the QMA reached the 0.5 mL (2 elutions of 0.25 mL each).

HPLC analysis was performed on a Shimadzu 20A/AT system with a Gabi detector (Elysia-Raytest; Straubenhardt, Germany) for the determination of the radiochemical and chemical purity as well as the chemical identity of [^{18}F]PSMA-1007 using Gina star software version 6.0 (Elysia-Raytest; Straubenhardt, Germany). For radioactivity detection an ACE 3 C18 150 x 1.5 mm (ACE) column was used. For the analysis, multi-step gradient was performed by using Water, 0.1% AcOH (solvent A) and acetonitrile, 0.1% AcOH (solvent B):20% B for the period 0-2.5 mins, 20 - 90% B for 9 minutes. Prior the quality control, reference standard of [^{19}F]PSMA-1007 was used for the determination for the retention time. A solution containing both [^{18}F]PSMA-1007 and the reference standard in equal volumes was also used after the HPLC run of the solution to be injected. GC (Shimadzu 17A, Lab solution 2.1 software; Kyoto, Japan) was used for the determination of residual solvents using a RESTEK RTX-624 (id 0.32 mm, length 1200 mm; Bellefonte, PA, USA) column. Radionuclidic identity was checked by half-life measurement via dose calibrator (M.E.D.-Medizintechnik Dresden GmbH, Dresden, Germany). Radionuclidic purity was determined using multi-channel analyser (Mucha star software version 6.0, Elysia -Raytest, Straubenhardt, Germany). Integrity of the filter used was assessed using ABT's Chemistry module. The final product was also tested for bacterial endotoxins (endosafe PTS nexgen with LAL- test, Charles River; Wilmington, Massachusetts, USA), pH (pH meter, HANNA instruments, Woonsocket, Rhode Island, USA), particulates and colourlessness. TLC (mini Gina detector and Gina star software, Elysia-Raytest; Straubenhardt, Germany) was performed using TLC Silica Gel 60 F_{254} Alu sheets (25 x 70mm). 1 μL of solution was spotted at 10 mm and run length was 50 mm. As development solvent, two combinations of acetonitrile and water were used (95:5 v/v ratio and 60/40 v/v ratio). Spot test was used for the quantification of the residual tetrabutylammonium ion (TBA). TLC Silica Gel 60 F_{254} Alu sheets and a mixture of MeOH:Ammonia (9:1 v/v ratio) were used as stationary and mobile phase respectively for the development of the plate. Reference standard of 0.75 M was used. The plate was then placed in an iodine chamber for two minutes. The reference standard had shown a dark brown spot at Rf of 0.3-0.4. Release of the radiopharmaceutical was performed by the responsible nuclear medicine physician after reviewing the tested parameters.

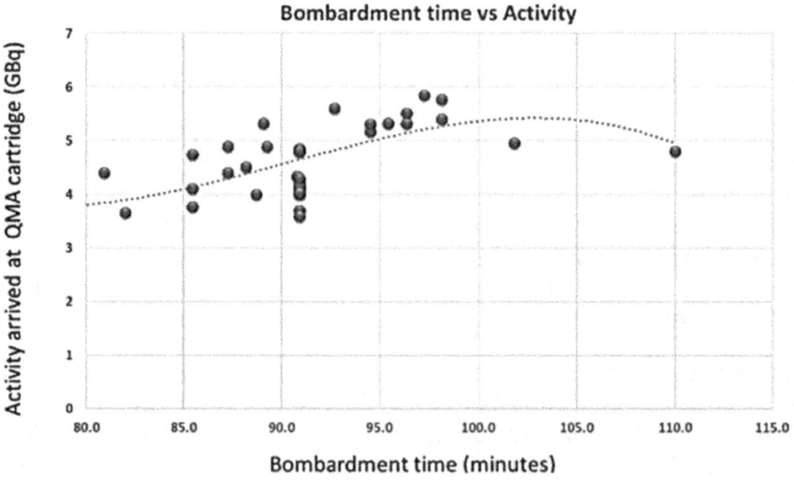

Figure 2 *Correlation between the time of bombardment and the activity arrived at the QMA cartridge.*

3 RESULTS

Time of bombardment and initial activity arriving at the QMA cartridges have been recorded. The results are depicted in Figure 2. The mean value of the radiochemical yield (decay corrected) achieved, was 57.0 ± 10.1%; with mean initial activity of 4.6 ± 0.9 GBq. The average activity of the final product measured was 2.1 ± 0.5 GBq. The range of concentrations of EtOH in the final product was found to be at the upper end of the limit; however, none of them exceeded the limit set by the European Pharmacopoeia (≤10% v/v). Regarding the rest of the residual solvents, (MeCN, DMSO and Acetone) their recorded values were lower than the limit of quantification. HPLC results have shown consistency on the activity graph and the radiochemical identity and chemical purity ≥95% in all 36 cases. Spiked solutions have shown successful results with respect to the reference standards (Figure 3). Mean production time is estimated at 40 mins; QC required additionally 25 mins. Average pH values were 6.8 ± 0.5. A radiochemical purity of more than 99% was recorded. R_f mean values of 0.37 (95:5 MeCN: H_2O v/v ratio) and 0.814 (60:40 MeCN:H_2O v/v ratio) were recorded using a radio TLC. The content of TBA in all final products was lower than the limit of quantification. The final volume of the product was 20 mL.

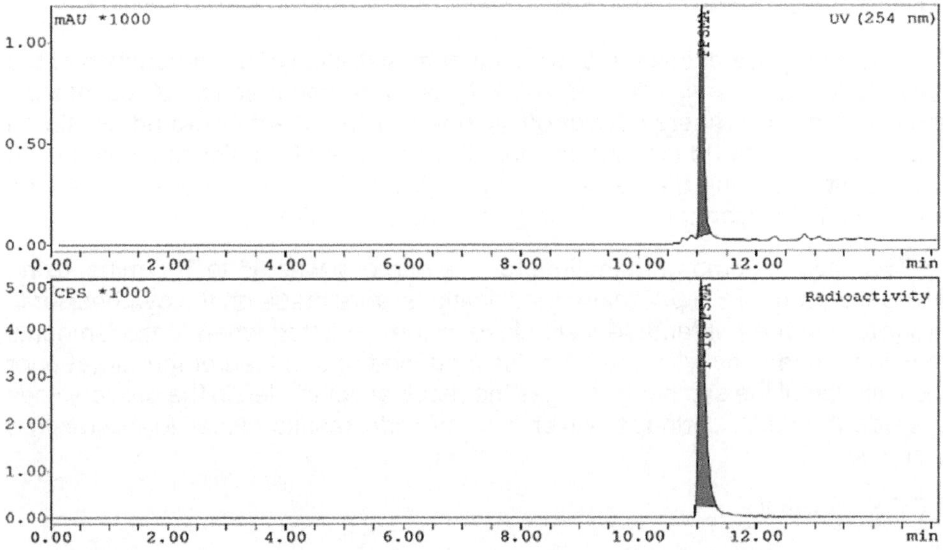

Figure 3 *Radio-HPLC spectrum of spiked [¹⁸F]PSMA1007. UV-Vis (top) and radioactive trace (bottom).*

4 PET IMAGING

Imaging was performed 180 minutes after intravenous administration of 3.8 ± 0.5 MBq/Kg*BW [^{18}F]PSMA-1007. The images were obtained from head to knees. Axial, sagittal and coronal PET reconstruction with and without attenuation correction were performed. After 36 successful syntheses, a total of 100 patients were able to undergo a PSMA-PET examination. A representative illustration of the examinations is provided in Figure 4.

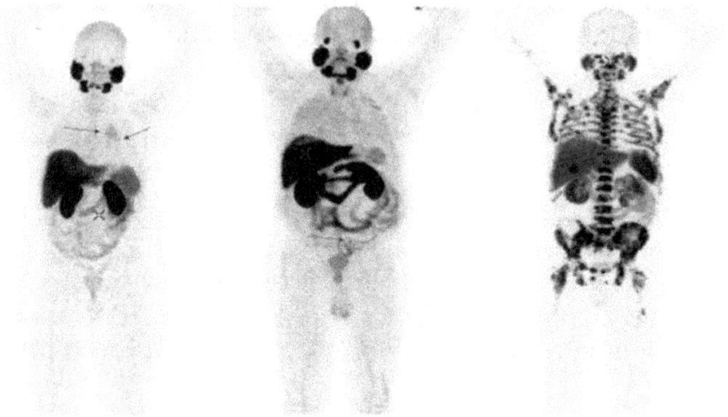

Figure 4 *Three different [^{18}F]PSMA-1007 images. 3.6 MBq/Kg*BW were used; Acquisition starts 180 minutes post injection; abdomen and pelvis 6 min/bp; rest 4 min/bp.*

5 DISCUSSION

The main purpose of this paper was the demonstration of the production of the new ABX radiotracer [^{18}F]PSMA-1007 using a commercial radiosynthesis module and a low energy cyclotron. Although 100 patients were administered and scanned after 36 production runs, the potential of administering/scanning more patients, using the same average activity of 2.1 GBq, purely depended on the PET/CT scanner's availability and characteristics.

The optimal bombardment time for this purpose was 95 to 100 mins, using a target current of 5.5 μA. Minor modifications were made on the synthesiser in order to reach a percentage yield close to the expected when larger amounts of volume and activity is used. The second load and unload of the target prior the initiation of the synthesis, purges the residual activity left in the delivery lines towards the QMA cartridge, which also provide results closer to the desired outcome.

6 CONCLUSION

To conclude, it has been shown that the production of [^{18}F]PSMA-1007 can be achieved with an average yield (d.c.) of 57.0 % using a low energy cyclotron. Since one batch of the radiotracer can be used for more than one patient, [^{18}F]PSMA-1007 can be considered as an alternative or replacement of the already existing and generator-based [^{68}Ga]PSMA compounds. Remote sites with an already-purchased low energy cyclotron can avoid the purchase of $^{68}Ge/^{68}Ga$ generator and use the already established production of ^{18}F. As a result [^{18}F]PSMA-1007 also proved economically viable for remote sites and local clinical environments.

CONFLICTS OF INTEREST

All authors are employees of German Oncology Centre, Limassol, Cyprus.

AUTHOR CONTRIBUTIONS

AF and MRP wrote the main manuscript text and prepared the Figures 1–3. AV wrote the PET imaging section and prepared Figure 4. All authors reviewed the study.

NOTES AND REFERENCES

1. R. Siegel, J. Ma, Z. Zou and A. Jemal, *CA Cancer J. Clin.*, 2014, **64**, 9–29.
2. L. A. Torre, F. Bray, R. L. Siegel, J. Ferlay, J. Lortet-tieulent and A. Jemal, *CA Cancer J. Clin.*, 2015, **65**, 87–108.
3. R. L. Siegel, K. D. Miller and A. Jemal, *CA Cancer J. Clin.*, 2018, **68**, 7–30.
4. C. B. Steele, J. Li, B. Huang and H. K. Weir, *Cancer*, 2017, **123**, 5160–5177.
5. F. L. Giesel, J. Cardinale, M. Schäfer, O. Neels, M. Benešová, W. Mier, U. Haberkorn, K. Kopka and C. Kratochwil, *Eur. J. Nucl. Med. Mol. Imaging*, 2016, **43**, 1929–1930.
6. C. Kesch, C. Kratochwil, W. Mier, K. Kopka and F. L. Giesel, *J. Nucl. Med.*, 2017, **58**, 687–688.
7. J. D. Young, V. Abbate, C. Imberti, L. K. Meszaros, M. T. Ma, S. Y. A. Terry, R. C. Hider, G. E. Mullen and P. J. Blower, *J. Nucl. Med.*, 2017, **58**, 1270–1277.
8. A. Afshar-Oromieh, T. Holland-Letz, F. L. Giesel, C. Kratochwil, W. Mier, S. Haufe, N. Debus, M. Eder, M. Eisenhut, M. Schäfer, O. Neels, M. Hohenfellner, K. Kopka, H. U. Kauczor, J. Debus and U. Haberkorn, *Eur. J. Nucl. Med. Mol. Imaging*, 2017, **44**, 1258–1268.
9. Z. Szabo, E. Mena, S. P. Rowe, D. Plyku, R. Nidal, M. A. Eisenberger, E. S. Antonarakis, H. Fan, R. F. Dannals, Y. Chen, R. C. Mease, M. Vranesic, A. Bhatnagar, G. Sgouros, S. Y. Cho and M. G. Pomper, *Mol. Imaging Biol.*, 2015, **17**, 565–574.
10. F. L. Giesel, B. Hadaschik, J. Cardinale, J. Radtke, M. Vinsensia, W. Lehnert, C. Kesch, Y. Tolstov, S. Singer, N. Grabe, S. Duensing, M. Schäfer, O. C. Neels, W. Mier, U. Haberkorn, K. Kopka and C. Kratochwil, *Eur. J. Nucl. Med. Mol. Imaging*, 2017, **44**, 678–688.
11. J. Cardinale, M. Schäfer, M. Benešová, U. Bauder-Wüst, K. Leotta, M. Eder, O. C. Neels, U. Haberkorn, F. L. Giesel and K. Kopka, *J. Nucl. Med.*, 2017, **58**, 425–431.
12. F. L. Giesel, C. Kesch, M. Yun, J. Cardinale, U. Haberkorn, K. Kopka, C. Kratochwil and B. A. Hadaschik, *Clin. Genitourin Cancer*, 2017, **15**, e497–e499.

13. ABT Molecular Imaging Inc., Technical Overview BG-75, 2015.
14. V. Awasthi, J. Watson, H. Gali, G. Matlock, A. McFarland, J. Bailey and A. Anzellotti, *Appl. Radiat. Isot.*, 2014, **89**, 167–175.
15. A. Anzellotti, J. Bailey, D. Ferguson, A. McFarland, P. Bochev, G. Andreev, V. Awasthi and C. Brown-Proctor, *J. Radioanal. Nucl. Chem.*, 2015, **305**, 387–401.
16. J. Cardinale, R. Martin, Y. Remde, M. Schäfer, A. Hienzsch, S. Hübner, A.-M. Zerges, H. Marx, R. Hesse, K. Weber, R. Smits, A. Hoepping, M. Müller, O. C. Neels and K. Kopka, *Pharmaceuticals*, 2017, **10**, DOI: 10.3390/ph10040077.

RAMAN ANALYSIS OF META-AUTUNITE

Victoria L. Frankland,*[a] Rachida Bance-Soualhi [a] and David Read [a, b]

[a] Department of Chemistry, University of Surrey, Guildford, GU2 7XH, UK.
[b] National Physical Laboratory, Teddington, TW11 0LW, UK.

* v.frankland@surrey.ac.uk

ABSTRACT

A sample of meta-autunite ($Ca(UO_2)_2(PO_4)_2{\cdot}6\text{-}8(H_2O)$) from a national reference collection was characterised by Raman spectroscopy as a representation of a potential spent nuclear fuel corrosion product. Raman spectra were collected at 457, 532, 633 and 785 nm; all exhibited some fluorescence effects, though to a lesser extent at 785 nm. The phosphate ($\nu_2(PO_4)^{3-}$, $\nu_3(PO_4)^{3-}$, $\nu_4(PO_4)^{3-}$) and uranyl ($\nu_1(UO_2)^{2+}$ and $\nu_2(UO_2)^{2+}$) features could be unambiguously assigned in the resolved 785 nm spectrum. The position of the $\nu_3(UO_2)^{2+}$ mode was predicted but not observed. The uranyl bond lengths and force constants were determined from the $\nu_1(UO_2)^{2+}$ dominant and shoulder peak, as 1.78 ± 0.01 and 1.79 ± 0.01 Å and 5.69 ± 0.08 and 5.29 ± 0.08 millidynes Å^{-1}, respectively.

1 INTRODUCTION

Meta-autunite ($Ca(UO_2)_2(PO_4)_2{\cdot}6\text{-}8H_2O$), autunite ($Ca(UO_2)_2(PO_4)_2{\cdot}10\text{–}12H_2O$) and other secondary uranyl alteration phases have environmental significance with regard to their potential to attenuate uranium and radionuclide mobility in groundwater. This is relevant to understanding the solubility and migration of radionuclides released during the corrosion of spent nuclear fuel (SNF), leaching of actinide-contaminated soils, and oxidation and leaching of uranium mine tailings [1-7]. Corrosion of SNF in groundwater is similar to the alteration of naturally-occurring uraninite (UO_2) mineralisation, which can provide a natural analogue insight into the corrosion process [2]. This typically proceeds from the development of ultra-thin films on the altered surface of the SNF, through colloidal particulates to aqueous solutions [8-12]. The physical persistence and chemical composition of these species is largely dependent on the surrounding solution, which can either enhance dissolution rates or promote the formation of passivating surfaces [13]. Early-formed colloidal products are typically characterised by uranyl oxide hydrates such as schoepite ($(UO_2)_8O_2(OH)_{12}$) [8]. Subsequent reaction with groundwater components (e.g. Ca, K, Na, silica, phosphate) forms a sequence of progressively more insoluble uranyl silicates such as soddyite, ($(UO_2)_2(SiO_4)(H_2O)_2$), and phosphates such as autunite [2]. Wherever feasible, theoretical prediction of leaching rates and the range of potential corrosion products formed needs to be verified by direct experimentation.

Approximately 250 natural uranium minerals are known, exhibiting a wide range of structural and chemical forms from which characterisation spectra of high quality reference minerals can be obtained. Among the most widespread of the natural uranium minerals is the autunite series of uranyl phosphate and arsenate hydrates with the general formula $M(UO_2)_2(XO_4)_2 \cdot x(H_2O)$ where *M* is the cation (e.g. Ca, Mg, Ba, Sr, Cu) and *X* is either P or As.

Laboratory investigations on synthetic uranyl oxide hydrates have revealed rapid transformation to autunite-structured compounds in the presence of phosphate solutions at room temperature [14]. Further work involving a metallic uranium precursor demonstrated the formation of a thin meta-autunite film [13, 15, 16] on the metal surface, which acted as a barrier to further dissolution or transformation. Thus, phosphate-bearing media have been proposed as potential backfill materials for radioactive waste repositories [17, 18] and for stabilization of radionuclides *in situ* [10-21]. Therefore, an understanding of the chemical characteristics of the autunite series of minerals is important when considering the management and disposal of uraniferous wastes and spent nuclear fuel [22].

Previous characterisation studies have been reported for meta-autunite samples using infrared (IR) [23-27] and Raman spectroscopy [25, 26, 28-30]. Difficulties were encountered with IR spectroscopy owing to overlap between the key $v_3(UO_2)^{2+}$, $v_1(PO_4)^{3-}$ and $v_3(PO_4)^{3-}$ features [25]. This overlap between modes is not observed in Raman spectroscopy but differences in peak positions and assignment are apparent [25, 28-30], which require further explanation.

In this study, a meta-autunite sample on loan from the British Geological Survey's Reference Collection (Ref. BGS 669/2) was characterised by Raman spectroscopy.

2 EXPERIMENTAL METHOD

The meta-autunite mineral sample originates from Mt. Spokane, Washington, US. Its composition was confirmed by powder X-ray diffraction and scanning electron microscopy with energy-dispersive X-ray microanalysis. For Raman characterisation, the sample was mounted and temporarily sealed into a groove between two quartz slides. A Renishaw InVia Reflex Raman Microscope was used with four lasers operating at 457 (Ar), 532 (Ar), 633 (HeNe) and 785 nm (diode laser), respectively. The Raman microscope [31] is equipped with a cooled charge-coupled detector (CCD) along with holographic notch filters and gratings designed for each laser wavelength. The apparatus is also fitted with an optical microscope (Leica, DMLM) with a range of objective lenses and a trinocular viewer equipped with a video camera enabling direct viewing of the sample. Daily calibrations performed using silicon ensured peak position accuracy of ± 1 cm^{-1}. Raman data were collected for each laser, between 50 and 4000 cm^{-1} from at least three different locations to confirm sample homogeneity.

3 RESULTS

3.1 Peak Resolution

Raman spectra for all of the laser wavelengths used are displayed in **Figure 1** over the full characterisation range. All spectra display a curved baseline, which increases with increasing wavenumbers, *v*. Negligible characterisation data can be obtained from the 457 and 532 nm spectra due to high fluorescence at these wavelengths. More Raman features are observed in the 633 nm spectrum, though they exhibit weak intensity. However, well defined features were observed across the whole *v* range using the 785 nm laser; therefore, this spectrum was resolved to identify the Raman modes present using the multiple peak fitting tool in Origin (Origin 2017, version 94E, OriginLab Corporation).

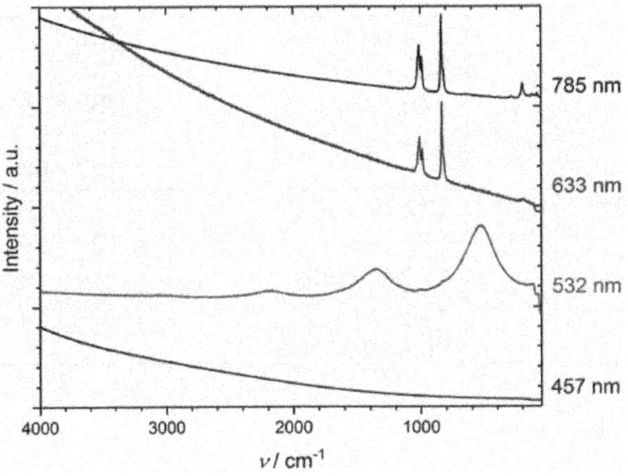

Figure 1 *Raman spectra collected for meta-autunite using 457, 532, 633 and 785 nm lasers.*

The resolved Raman spectrum is shown in **Figure 2** and is sub-divided into four regions: 1200 – 900, 900 – 750, 700 – 350 and 350 – 50 cm^{-1}. Two water features were expected for meta-autunite at high *v*; the first being the hydroxide vibration, *v*(OH), between 3800 – 3200 cm^{-1} [32-34] and the second a water-bending mode, $\delta(H_2O)$, at 1600 – 1750 cm^{-1} [32-34]. Neither was detected in the present study, due to fluorescence effects masking these weak intensity features. The lack of observation of the water features in this study could have possibly been due to differences in the apparatus settings, especially if the other studies used a bleaching mode which delays the collection of a Raman spectrum until after the termination of the laser pulse enabling time for the fluorescence to decay to lower intensities.

Table 1 *Summary of Raman data for meta-autunite mineral sample. Intensity described as strong, s; medium, m; weak, w; and very weak, vw; shoulder peaks as sh.*

Assignment	***v* (this study) / cm^{-1}**	***v* (literature) / cm^{-1}**
$v_3(PO_4)^{3-}$	1107 (vw, b) 1034 (w) 1020 (m, sh) 1009 (m) 990 (m) 967 (vw)	1093 [25, 28], 1111 [29] 1033 [28] 1018 [25, 28] 1007 [25, 28], 1008 [29] 988 [29], 989 [25, 28]
$v_1(UO_2)^{2+}$	836 (s) 821 (m, sh)	833 [25, 28, 29] 816 [29], 818 [25, 28]
$v_4(PO_4)^{3-}$	659 (vw) 635 (vw) 556 (vw)	643 [25, 28] 635 [29] 507 [25, 28]
$v_2(PO_4)^{3-}$	455 (vw) 390 (vw)	453 [25, 28] 387 [25, 28]
$v_2(UO_2)^{2+}$	306 (vw, sh) 289 (vw) 276 (vw, sh)	303 [30] 285 [30] 263 [25, 28] 239 [30] 222 [25, 28]
Lattice vibrations	222 (vw, sh) 197 (w) 176 (vw, sh) 121 (vw) 80 (vw)	190 [25, 28]

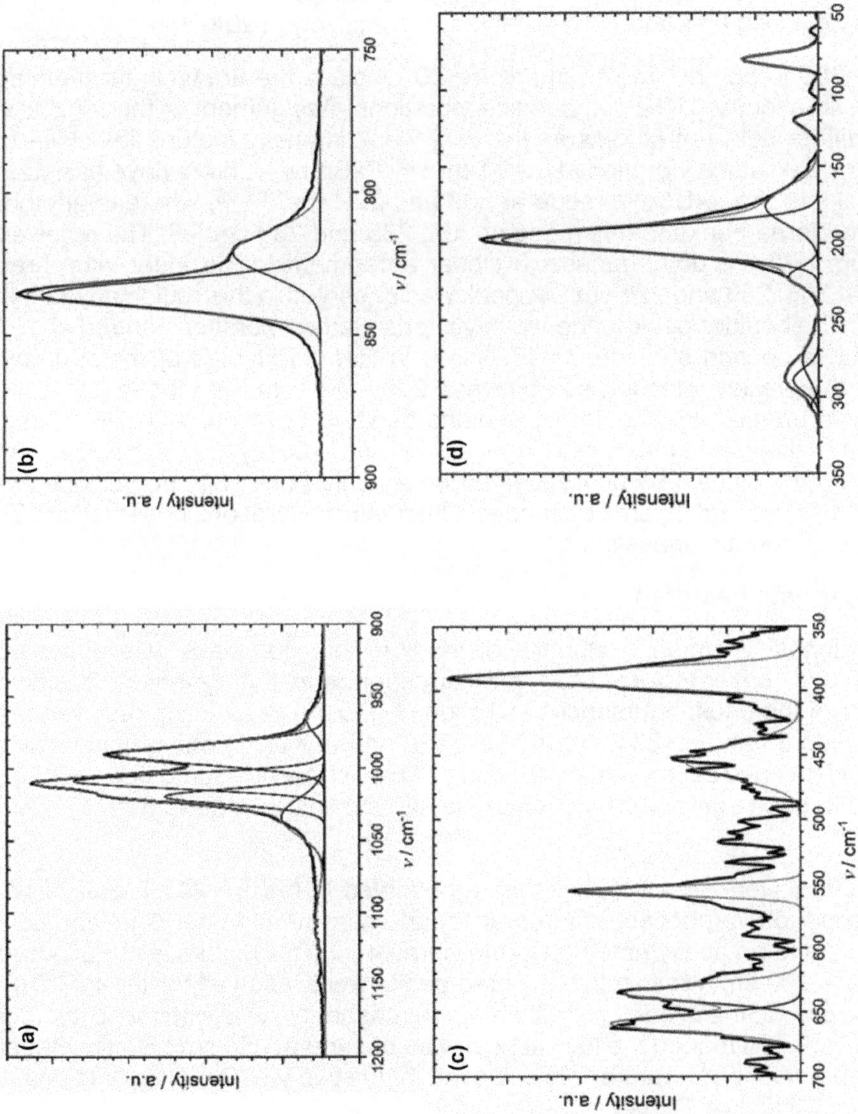

Figure 2 *Peak fitting results for the 785 nm spectrum.*

3.2 Uranyl Features

The most intense Raman peak observed is the uranyl symmetric stretch vibration, $v_1(UO_2)^{2+}$. This feature is centred at ~ 835 cm^{-1} and was resolved into two peaks at 836 (dominant) and 821 cm^{-1} (shoulder, **Figure 2B**). Similar results have been reported for meta-autunite samples from France [25,28] and Hungary [29] using 633 and 785 nm lasers, respectively (**Table 1**).

The 350 to 50 cm^{-1} region in **Figure 2D** displays the uranyl antisymmetric stretch vibration, $v_2(UO_2)^{2+}$ and lattice vibrations. Assignment of the $v_2(UO_2)^{2+}$ mode differs between studies as the very weak intensity feature lies close to another weak lattice vibration at ~ 190 cm^{-1} [25, 28]. Some workers have assigned two peaks to the $v_2(UO_2)^{2+}$ mode at 263 and 222 cm^{-1} [25, 28], whereas another attributed three peaks to this mode at 303, 285 and 239 cm^{-1} [29]. The resolved spectrum in this study suggests a closer agreement to the latter with three peaks at 306, 289 and 276 cm^{-1}. A peak was observed in this region at 222 cm^{-1} but it is a shoulder peak to the weakly intense lattice vibration centred at 197 cm^{-1} with a second shoulder at 176 cm^{-1}. In Raman studies of metal uranyl oxides, the weakly intense peak between 200 – 190 cm^{-1} has been attributed to either a uranium-hydroxide out of plane bending vibration, $\gamma(U_3(OH)_3)$, or a uranium-hydroxide in plane bending vibration, $\delta(U_3(OH)_3)$, [36, 37]. The 222, 197 and 176 cm^{-1} lattice vibration observed here could potentially be assigned to one of these uranium hydroxide modes. Other lattice vibrations observed at 121 and 80 cm^{-1} remain unassigned.

3.3 Phosphate Features

In the highest v region, a doublet of medium intense peaks was observed around 1000 cm^{-1} (**Figure 2A**). This corresponds to the symmetric bending vibration of the phosphate functional group, $v_3(PO_4)^{3-}$. Peak fitting resolved the doublet into 5 bands (1034, 1020, 1009, 990 and 967 cm^{-1}) with a very weakly intense sixth band situated at higher v (1107 cm^{-1}). The position of the $v_3(PO_4)^{3-}$ mode is in good agreement with other Raman studies on meta-autunite [25, 28, 29] (**Table 1**).

Five very weak intense peaks were observed in **Figure 2C**. These peaks correspond to two phosphate ion modes: an antisymmetric bending vibration, $v_4(PO_4)^{3-}$, and an antisymmetric stretch vibration, $v_2(PO_4)^{3-}$, between 680 - 530 and 500 – 300 cm^{-1}, respectively. Three peaks were resolved for the $v_4(PO_4)^{3-}$ mode at 659, 635 and 556 cm^{-1} which agree reasonably well with those in other studies [25, 28, 29], although the lowest peak was observed ~ 50 cm^{-1} higher. Better agreement with the literature was achieved for the two $v_2(PO_4)^{3-}$ resolved peaks at 455 and 390 cm^{-1} (**Table 1**).

4 DISCUSSION

A key Raman transition relevant to spent nuclear fuel (SNF) corrosion product studies is uranyl ion (UO_2^{2+}) vibration, which usually displays a strong symmetric stretch, $v_1(UO_2)^{2+}$, situated typically between 800 and 890 cm^{-1} [38].

This mode is sensitive to factors such as the number and type of equatorial bonding ligands present. Electrons donated by these equatorial ligands to UO_2^{2+} enter the ion's lowest unoccupied levels, δ_u and ϕ_u, which are derived solely from U 5f orbitals [39, 41]. The more electrons donated to UO_2^{2+}, the greater the repulsion of bonded O atoms leading to the O-U-O bonds becoming weaker and longer in length. This decrease in vibrational frequency results in the Raman $\nu_1(UO_2)^{2+}$ mode shifting to higher wavenumbers. A more extreme example is $CaUO_4$ where the ν_1(O-U-O) Raman mode is observed at 696 cm^{-1} [41].

A relationship between the ν_1(O-U-O) Raman mode (in cm^{-1}) and the U-O bond length, $d_{U\text{-}O}$, (in Å) was developed by Bartlett and Cooney [33]. It is given by **Equation 1** and holds well for all uranyl minerals.

$$d_{U-O} = 106.5[\nu_1(UO_2)^{2+}]^{-2/3} + 0.575 \qquad (1)$$

The values of $d_{U\text{-}O}$ for the two $\nu_1(UO_2)^{2+}$ peaks at 836 and 821 cm^{-1} are 1.78 ± 0.01 and 1.79 ± 0.01 Å, respectively, where the error was calculated from the variation in the $\nu_1(UO_2)^{2+}$ peak position between repeats (± 3 cm^{-1}). The two $d_{U\text{-}O}$ values are within experimental error. An XRD analysis by Locock and Burns [5] obtained $d_{U\text{-}O}$ values for autunite of 1.784 and 1.794 Å, which are close to those obtained in this study and also agree with the observations of two resolved peaks in the $\nu_1(UO_2)^{2+}$ feature. Although XRD is the commonly preferred technique for obtaining $d_{U\text{-}O}$ values, it is limited to well crystalline phase samples only (such as meta-autunite). Raman characterisation can be applied to solids (both crystalline and amorphous), colloids and aqueous solutions. Thus, $d_{U\text{-}O}$ can be extracted from a greater range of possible uranium(VI)-bearing SNF corrosion products which is the long-term aim of this ongoing characterisation research.

Bartlett and Cooney [42] also derived a second formula linking the symmetric bending vibration, $\nu_3(UO_2)^{2+}$, to $d_{U\text{-}O}$ (**Equation 2**). This second equation is more commonly associated with IR characterisation as the $\nu_3(UO_2)^{2+}$ is IR strong and Raman weak. However, this mode has been observed in Raman studies of other uranyl phosphate species such as meta-torbernite [25, 28, 30, 43, 44]. From the known $d_{U\text{-}O}$ for this sample, the position of the $\nu_3(UO_2)^{2+}$ peaks were predicted as 913 ± 4 and 893 ± 4 cm^{-1}, respectively. This prediction agrees well with IR observations of meta-autunite and autunite samples [23, 24, 45]. However, no features corresponding to the $\nu_3(UO_2)^{2+}$ mode were detected in this Raman study (**Table 1**) nor has the $\nu_3(UO_2)^{2+}$ mode been observed in other meta-autunite Raman studies [25, 28, 29].

$$d_{U-O} = 92.42[\nu_3(UO_2)^{2+}]^{-2/3} + 0.804 \qquad (2)$$

Another value that can be obtained from $d_{U\text{-}O}$ is the U-O bond force constant, $k_{U\text{-}O}$, using Badger's rule (**Equation 3**) [46, 47]:

$$k_{U-O} = [1.08/(d_{U-O} - 1.17)]^3 \qquad (3)$$

where $k_{U\text{-}O}$ is in units of millidyne $Å^{-1}$ (where 1 millidyne $Å^{-1}$ equals 100 N m^{-1}). In this study, $k_{U\text{-}O}$ for the two $v_1(UO_2)^{2+}$ peaks are 5.69 ± 0.08 and 5.29 ± 0.08 millidynes $Å^{-1}$, respectively. These values are similar to those calculated for other uranyl phosphate minerals [30] but, to the best of our knowledge, have not been published for meta-autunite. Applications of the $k_{U\text{-}O}$ values for meta-autunite is expected to be useful to those modelling the crystal structure of meta-autunite and for exploring the formation pathways of uranium(VI)-bearing SNF corrosion products.

5 CONCULSIONS

Chemical composition of a meta-autunite ($Ca(UO_2)_2(PO_4)_2{\cdot}6\text{-}8(H_2O)$) sample from a national reference collection was obtained by Raman spectroscopy. Fluorescence effects were observed for all of the lasers used (457, 532, 633 and 785 nm) but to a lesser extent for the 785 nm spectrum. Resolving the latter revealed: two uranyl modes $v_1(UO_2)^{2+}$ (836 and 821 cm^{-1}) and $v_2(UO_2)^{2+}$ (306, 289 and 276 cm^{-1}); three phosphate modes, assigned as $v_3(PO_4)^{3-}$ (1107, 1034, 1020, 1009, 990 and 967 cm^{-1}), $v_4(PO_4)^{3-}$ (659, 635 and 556 cm^{-1}) and $v_2(PO_4)^{3-}$ (455 and 390cm^{-1}) and lattice vibrations at 222, 197, 176, 121 and 80 cm^{-1}. Predictions were made for the location of the weakly intense $v_3(UO_2)^{2+}$ mode as 913 ± 2 and 893 ± 1 cm^{-1} but this feature was not observed. The two uranyl U-O bond lengths and bond force constants were calculated as 1.78 ± 0.01 and 1.79 ± 0.01 Å and 5.69 ± 0.08 and 5.29 ± 0.08 millidynes $Å^{-1}$, respectively. Characterisation of this and other minerals by Raman is underway along with other techniques such as time-resolved laser fluorescence spectroscopy in order to develop a reference database for potential SNF corrosion products.

CONFLICTS OF INTEREST

There are no conflicts to declare.

ACKNOWLEDGEMENTS

This work was supported by funding from the Engineering & Physical Science Research Council (project number EP/L014041/1 – DISTINCTIVE). The authors wish to thank Kay Green of the British Geological Survey for the loan of the reference mineral sample. Acknowledgements are also given to John-William Brown, Sarah Heisig and Carol Crean for advice and support with the Raman apparatus.

REFERENCES

1. W. H. Hocking, A. M. Duclos, and L. H. Johnson, *J. Nucl. Mater.*, 1994, **209**, 1.
2. P. C. Burns, *Amer. Mineral.*, 1997, **82**, 1176.
3. J. L. Jerden Jr., and A. K. Sinha, *Appl. Geochem.*, 2003. **18**, 823.
4. J. L. Jerden Jr., A. K. Sinha, and L. Zelazny, *Chem. Geol.*, 2003, **199**, 129.
5. A. J. Locock and P. Burns, *Amer. Mineral.*, 2003, **88**, 240.

6. A. J. Locock, P. C. Burns and T. M. Flynn, *Canadian Mineral.*, 2005, **43**, 721.
7. J. G. Catalano, J. P. McKinley, J. M. Zachara, S. M. Heald, S. C. Smith, and G. E. Brown, Jr, *Environ. Sci. Technol.*, 2006. **40**, 2517.
8. P. A. Finn, E. C. Buck, M. Gong, J. C. Hoh, J. W. Emery, L. D. Hafenrichter, and J. K. Bates, *Fourth International Conference on the Chemistry and Migration Behavior of the Actinides and Fission Products in the Geosphere MIGRATION '93 Charleston, SC December 12-17, 1993.*
9. M. D. Kaminski, N. M. Dimitrijevic, C. J. Mertz and M. M. Goldberg, *J. Nucl. Mater.*, 2005, **347**, 77.
10. V. Moulin and G. Ouzounian, *Appl. Geochem.*, 1992, **7**, 179.
11. X. Wang, Y. Sun, A. Alsaedi, T. Hayat and X. Wang, *Chem. Eng. J.*, 2015, **264**, 570.
12. A. Kowal-Fouchard, R. Drot, E. Simoni and J. Ehrhardt, *Environ. Sci. Technol.*, 2004, **38**, 1399.
13. N. Baumann, T. Arnold, H. Foerstendorf and D. Read, *Environ. Sci. Technol.*, 2008, **42**, 8266.
14. A. Sowler, S. Clark and R. Fjeld, *Radiochim. Acta*, 1996, **74**, 45.
15. T. Murakami, T. Ohnuki, H. Isobe and T. Sato, *Amer. Mineral.*, 1997, **82**, 888.
16. N. Baumann, T. Arnold, G. Geipel, E. Trueman, S. Black and D. Read, *Sci. Total Environ.*, 2006, **366**, 905.
17. R. Gauglitz, M. Holterdorf, W. Franke and G. Marx, *Radiochim. Acta*, 1992, **58-9**, 253.
18. X. Li, L. Zhou, G. Zhao and Z. Liu, *Gospod. Surowcami. Min.*, 2008, **24**, 225.
19. X. Zhang, T. Jiang, C. Xie, Y. Peng, M. Li and Y. Zhong, *Water Air Soil Poll.*, 2018, **229**.
20. H. Wen, Z. Pan, D. Giammar and L. Li, *Environ. Sci. Technol.*, 2018, **52**, 5841.
21. L. Lammers, H. Rasmussen, D. Adilman, J. de Lemos, P. Zeeb, D. Larson and A. Quicksall, *Appl. Geochem.*, 2017, **84**, 105.
22. R. Baker, *Coordin. Chem. Rev.*, 2014, **266**, 123.
23. J. Cejka, A. Muck and J. Cejka, *Phys. Chem. Miner.*, 1984, **11**, 172.
24. B. Lazic, V. Kahlenberg, P. Vulic, L. Pesic and R. Dimitrijevic, *Neues Jb. Miner. Abh.*, 2009, **186**, 333.
25. R. Frost, *Spectrochim. Acta A*, 2004, **60**, 1469.
26. B. Lafuente, R. Downs, H. Yang and N. Stone, *The power of databases: the RRUFF project*, Berlin, Germany, 2015.
27. V. Bermanec, R. Wegner, G. Kniewald, B. Rakvin, L. Palinkas, M. Rajic, N. Tomasic and K. Furic, *Neues Jb. Miner. Abh.*, 2005, **181**, 27.
28. R. Frost and M. Weier, *Spectrochim. Acta A*, 2004, **60**, 2399.
29. E. Stefaniak, A. Alsecz, R. Frost, Z. Mathe, I. Sajo, S. Torok, A. Worobiec and R. Van Grieken, *J. Hazard. Mater.*, 2009, **168**, 416.

30. E. Faulques, N. Kalashnyk, F. Massuyeau and D. Perry, *RSC Adv.*, 2015, **5**, 71219.
31. W. Lee, C. Crean, J. Varcoe and R. Bance-Soulahi, *RSC Adv.*, 2017, **7**, 47726.
32. B. Kolesov and C. Geiger, *Amer. Mineral.*, 2002, **87**, 1426.
33. N. Zotov and H. Keppler, *Amer. Mineral.*, 1998, **83**, 823.
34. S. Pershin, *Opt. Spectrosc.*, 2005, **98**, 543.
35. A. Bell, L. Hecht and L. Barron, *J. Amer. Chem. Soc.*, 1997, **119**, 6006.
36. R. Frost, J. Cejka and M. Weier, *J. Raman Spectrosc.*, 2007, **38**, 460.
37. R. Frost, M. Dickfos and J. Cejka, *J. Raman Spectrosc.*, 2008, **39**, 1158.
38. R. Driscoll, D. Wolverson, J. Mitchels, J. Skelton, S. Parker, M. Molinari, I. Khan, D. Geeson and G. Allen, *RSC Adv.*, 2014, **4**, 59137.
39. E. Koglin, H. Schenk and K. Schwochau, *Spectrochim. Acta A*, 1979, **35**, 641.
40. S. McGlynn, W. Neely and J. Smith, *J. Chem. Phys.*, 1961, **35**, 105.
41. M. Liegeois-Duyckaerts, *Spectrochim. Acta A*, 1977, **33**, 709.
42. J. Bartlett and R. Cooney, *J. Mol. Struct.*, 1989, **193**, 295.
43. N. Sanchez-Pastor, A. Pinto, J. Astilleros, L. Fernandez-Diaz and M. Goncalves, *Spectrochim. Acta A*, 2013, **113**, 196.
44. F. Cretaz, S. Szenknect, N. Clavier, P. Vitorge, A. Mesbah, M. Descostes, C. Poinssot and N. Dacheux, *J. Nucl. Mater.*, 2013, **442**, 195.
45. J. Cejka, J. Cejka and A. Muck, *Thermochim. Acta*, 1985, **86**, 387.
46. R. Badger, *Phys. Rev.*, 1935, **3**, 284.
47. L. Jones, *Spectrochim. Acta*, 1959, **15**, 409.

RAPID DETERMINATION OF SR-90 IN ENVIRONMENTAL MATRICES BY SPE-ICP-MS FOR EMERGENCY MONITORING

Kenso Fujiwara*‡a, Kayo Yanagisawa‡ b and Kazuki Iijima [a]

[a] 10-2, Fukasaku, Miharu-machi, Tamura-gun, Fukushima 963-7700, Fukushima Radiation Measurement Group Fukushima Environmental Safety Center Japan Atomic Energy Agency. [b] 1-3, Nodzuki, Obuchi, Rokkasho-mura, Aomori 039-3212, Japan Nuclear Fuel Chemical Analysis Co. Ltd.

*fujiwara.kenso@jaea.go.jp

ABSTRACT

In order to evaluate the migration and transfer of radionuclides in the environment, not only Strontium-90 (Sr-90) in water but also those in soil and biological samples should be measured. For rapid quantification of Sr-90, a new method by means of online solid phase extraction coupled with inductively coupled plasma mass spectrometry (SPE-ICP-MS) has been developed. In this method, it is unavoidable to eliminate the interference by polyatomic ion induced by isotopes and matrices in fishes. In this study, SPE-ICP-MS method was applied to the measurement of Sr-90 in fishes, and the influence of the Sr isotopes and coexisting ions such as Ca element was evaluated. The detection limit (3σ) was evaluated as 0.35Bq/L (740Bq/kg), which was applicable level to WHO guidelines for drinking water quality.

1 INTRODUCTION

Since Sr-90 is one of the high fission yield and hazardous fission products, accurate and low-level detection of Sr-90 is essential for environmental samples. Especially, in case of nuclear power plant accidents, rapid and precise assessment of Sr-90 for environmental monitoring and health physics

---

‡ These authors contributed equally to this work.

is required since Sr-90 is considered to be accumulated in the body related to distribution of calcium (Ca). And in order to evaluate the influence of environmental migration of radionuclides, it is important to confirm the chemical condition of Sr-90 in the soil, fish and/or plant. In the case of β counting techniques of Sr-90, chemical separation of Y-90 is required. It is taken time for radiation equilibrium or growth of Y-90.

For rapid analysis method of Sr-90 in environmental water, SPE-ICP-MS system was established. SPE-ICP-MS system is measured easily quantify ultratrace isotopes including Sr-90, combined with ICP-QMS with an oxygen reaction cell, ultrasonic nebuliser and online solid-phase extraction (SPE) system.[1] This method is no requirement of radiation equilibrium compared with official method based on β ray measurement for separation from Y-90 by using Sr resin. Since analysis time of SPE-ICP-MS will be expected to shorten, this measurement can be cited as an option for monitoring. However, in the case of a solid sample, since the matrix ion concentration is higher than that of the water sample, it is necessary to eliminate and/or evaluate the effects of Sr isotopes and polyatomic by element in matrices. Since measuring of Sr-90 in water and other samples is essential to understand dynamics of radionuclides in the environment, in recent years, several methods to correct fluctuated the signal intensity of the target element and recovery rate due to introduction of large amounts of matrix etc. was developed.[2,3,4] However, in the case of fish by SPE-ICP-MS system, it is necessary to obtain recovery ratio by using intensity of stable isotope and confirm polyatomic interference. In the case of recovery ratio, it is reported that Sr-88 and Sr-86 were able to use for recovery ratio,[2,4] however, it is difficult to use for high concentration of Sr-88 or Sr-86 in matrix of fish. And for confirmation of polyatomic interference, concentration of element in matrixes measured and it is necessary to evaluate the influence of polyatomic species. In this study, at first, element of fish sample is measured and at ICP-MS measurement, the constituent elements of the polyatomic species affecting Sr were evaluated. Next, in order to confirm the recovery ratio is able to calculated by signal of Sr-84, constituent elements of polyatomic species and Sr-84 were measured by ICP-MS and influence of polyatomic species evaluated. Last, for environmental analysis especially food monitoring, concentrations of Sr-90 in the fishes spiked were measured by SPE-ICP-MS system and the influence of the matrix such as isotope of Sr and polyatomic ion was confirmed.

2 MEASUREMENT

2.1 Inorganic elements in fish

For confirmation of treatment method, elements of NIES CRM No.27[5] was measured by ICP-AES (ICPE-9000, Shimadzu). And for evaluating the influence of matrix by spike test of Sr-90, elements of several fish were measured by the same method. Fish samples were collected from the coast of Fukushima. Fish samples were kept in a freezer as preliminary freezing at -55 degrees overnight after filleted the bone and edible parts. These samples were freeze dried by Freeze dryer (Alpha 1-2 LD plus, Martin Christ GmbH.). The dried fish samples and NIES sample (edible parts and NIES sample: approximately 0.5g, bone parts: approximately 0.1g) were digested with 8.0ml 8.0M HNO_3 in a microwave oven (ETHOS One, Milestone General K.K.). After digested solutions were added ultrapure water to 50ml, diluted 100 fold to give a 20% nitric acid solution. Inorganic elements of sample solutions filtered by using 0.45μm PTFE filter were measured by ICP-AES.

2.2 Sr-90 spike test and matrix effect for system validation

For confirmation of polyatomic interference (e.g. ArCa, Ca_2 and SO_3),[6,7,8] 3 sample solutions (Sr:1ppb, Sr:1ppb + Ca:10ppm and Sr:1ppb + S:10ppm) were adjusted based on the concentration ratio of result of (1) by standard solutions with 20v/v% HNO_3. Concentration of Sr-84 in sample was measured by SPE-ICP-MS system (NexION300X, PerkinElmer).[3,4]

For Sr-90 spike test for system validation and calculation of detection limit, freeze-dried sample of edible part of brown hakelings were digested with 8.0ml 8.0M HNO_3 in a microwave oven. Sample solutions filled up to 50mL with ultrapure water. After filtration by using 0.45μm PTFE filter, sample solutions were adjusted by changing dilution ratio by adding HNO3 to give 20v/v% and spiked Sr-90 concentration (Table 2). Sr-90 of sample solutions was measured by SPE-ICP-MS system.

3 RESULTS AND DISCUSSION

3.1 Inorganic elements in fish

Table 1 shows concentration of inorganic element in brown hakeling, horse mackerel, whitebait and NIES CRM No27 sample. About NIES CRM No27 sample, since there was no difference between certified value and measured value, it is considered that there is no problem in processing after freeze drying.

Comparing the concentration of strontium and other element, in bone part, calcium concentration was found to be about 100 to 1000 times higher and in edible part, potassium concentration was found to be about 100 to 1000 times higher. In the case of measuring by using ICP-MS, samples must be diluted 100 to 1000 times in order to avoid several problems such as clogging of interface ones and ion deflector, and contamination of cells. In the case of measuring Sr-84, it is necessary to consider the influence of Ca at concentration 1000 times higher than that of Sr.

Table 1 *Contents of inorganic elements in brown hakeling.*

Sample			Water [%]	Na [g kg^{-1}]	Mg [g kg^{-1}]	K [g kg^{-1}]	Ca [g kg^{-1}]
Horse mackerel, n=3	Edible part	Mean.	69.8 ± 3.6	3.19 ± 0.76	2.04 ± 0.20	22.3 ± 3.4	1.90 ± 1.00
		(Min.~Max.)	(67.0 ~ 73.9)	(2.64 ~ 4.06)	(1.85 ~ 2.26)	(19.1 ~ 25.9)	(0.88 ~ 2.88)
	Bones*	Mean.	46.9 ± 7.3	2.93 ± 0.41	1.22 ± 0.24	3.76 ± 0.81	157.6 ± 2.67
		(Min.~Max.)	(41.5 ~ 55.2)	(2.67 ~ 3.40)	(1.05 ~ 1.49)	(2.90 ~ 4.51)	(154.6 ~ 159.6)
Brown hakeling, n=3	Edible part	Mean.	78.7 ± 0.6	3.23 ± 0.17	1.30 ± 0.05	14.9 ± 0.4	1.42 ± 1.18
		(Min.~Max.)	(77.9 ~ 79.2)	(3.00 ~ 3.46)	(1.22 ~ 1.37)	(14.3 ~ 15.4)	(0.85 ~ 4.53)
	Bones*	Mean.	58.6 ± 1.6	3.16 ± 0.31	3.01 ± 0.22	5.06 ± 0.25	168.1 ± 15.9
		(Min.~Max.)	(57.1 ~ 60.3)	(2.88 ~ 3.49)	(2.84 ~ 3.26)	(4.85 ~ 5.34)	(155.3 ~ 186.0)
Whitebait, n=4	Whole	Mean.	83.5 ± 0.4	19.7 ± 3.2	3.08 ± 0.49	22.3 ± 0.6	7.48 ± 0.69
		(Min.~Max.)	(82.8 ~ 84.0)	(16.2 ~ 22.7)	(2.75 ~ 3.79)	(21.5 ~ 23.0)	(6.87 ~ 8.08)
Octpus, n=1	Whole		81.1	32.0	2.58	5.10	1.01
NIES CRM No.27	Measured			10.3 ± 0.5	0.607 ± 0.026	5.44 ± 0.18	1.32 ± 0.07
	Certified			10.0 ± 0.4	0.576 ± 0.012	5.50 ± 0.15	1.25 ± 0.04
	Recovery (%)			103	105	99	106

Sample			Sr [mg kg^{-1}]	Ba [mg kg^{-1}]	Cu [mg kg^{-1}]	Mn [mg kg^{-1}]	Zn [mg kg^{-1}]
Horse mackerel, n=3	Edible part	Mean.	25.7 ± 11.0	2.5 ± 2.4	6.4 ± 0.8	1.3 ± 0.4	147.2 ± 15.5
		(Min.~Max.)	(13.7 ~ 35.2)	(N.D. ~ 4.1)	(5.6 ~ 7.2)	(0.85 ~ 1.59)	(133.0 ~ 163.7)
	Bones*	Mean.	797.3 ± 81.2	40.1 ± 4.2	N.D.	4.8 ± 0.8	58.9 ± 12.1
		(Min.~Max.)	(707.1 ~ 864.7)	(36.3 ~ 44.6)		(4.3 ~ 5.7)	(45.6 ~ 69.2)
Brown hakeling, n=3	Edible part	Mean.	5.0 ± 4.4	0.6 ± 0.5	1.6 ± 0.4	0.5 ± 0.2	23.0 ± 1.8
		(Min.~Max.)	(2.1 ~ 16.5)	(N.D. ~ 1.3)	(1.1 ~ 2.0)	(0.3 ~ 1.0)	(21.5 ~ 26.9)
	Bones*	Mean.	399.1 ± 28.5	18.5 ± 7.1	N.D.	11.9 ± 4.0	65.4 ± 4.9
		(Min.~Max.)	(375.4 ~ 430.7)	(12.5 ~ 26.3)		(8.6 ~ 16.3)	(61.8 ~ 71.0)
Whitebait, n=4	Whole	Mean.	26.8 ± 3.0	1.1 ± 0.5	1.0 ± 0.1	2.4 ± 0.2	44.4 ± 4.1
		(Min.~Max.)	(24.2 ~ 30.7)	(0.5 ~ 1.6)	(0.8 ~ 1.2)	(2.1 ~ 2.6)	(38.7 ~ 48.1)
Octpus, n=1	Whole		17.2	1.2	16.7	0.9	101.2
NIES CRM No.27	Measured		4.8 ± 0.2	1.0 ± 0.8	3.0 ± 0.3	8.7 ± 0.4	20.0 ± 0.9
	Certified		4.9 ± 0.2	1.1 ± 0.1	2.8 ± 0.1	8.9 ± 0.2	20.9 ± 0.9
	Recovery (%)		98	91	107	98	96

*Bones consists mainly of the central vertebra

3.2 Sr-90 spike test and matrix effect for system validation

Figure 1 shows the schematic of SPE-ICP-MS system.[1] Sample solution and standard solution are divided into two flow lines in this system. In the case of #2, One flows to the Column for concentration, but the other flows to ICP-MS directly. Sr-84 is measured by split-flow line and Sr element was concentrated by column. In the case of #1, Sr element in the column was eluted with water and Sr-84 and 90 was measured. On both lines, In-115 was added as an internal standard. The concentration was measured by calibration curve method by using standard solution of Sr stable isotopes.[3,4] In the case of sample solution and standard solution, intensity of split flow line is the same, but intensity of column line is will (or might) lower in the sample solution because the recovery ratio is affected by coexisting matrices.

Measuring of the signal intensities from each both lines make it possible to calculate the relative recovery ratio. Relative recovery percentage (R%) is calculated intensity of the sample solution when intensity of standard solution is set to 100% was calculated by method.[4]

Figure 2 shows signal profiles of Sr-84. Since intensity of Sr-84 of split-flow line and column line measured clearly, the concentrations of Sr-90 were measured by SPE-ICP-MS system and calculated by using R% of Sr-84.[4]

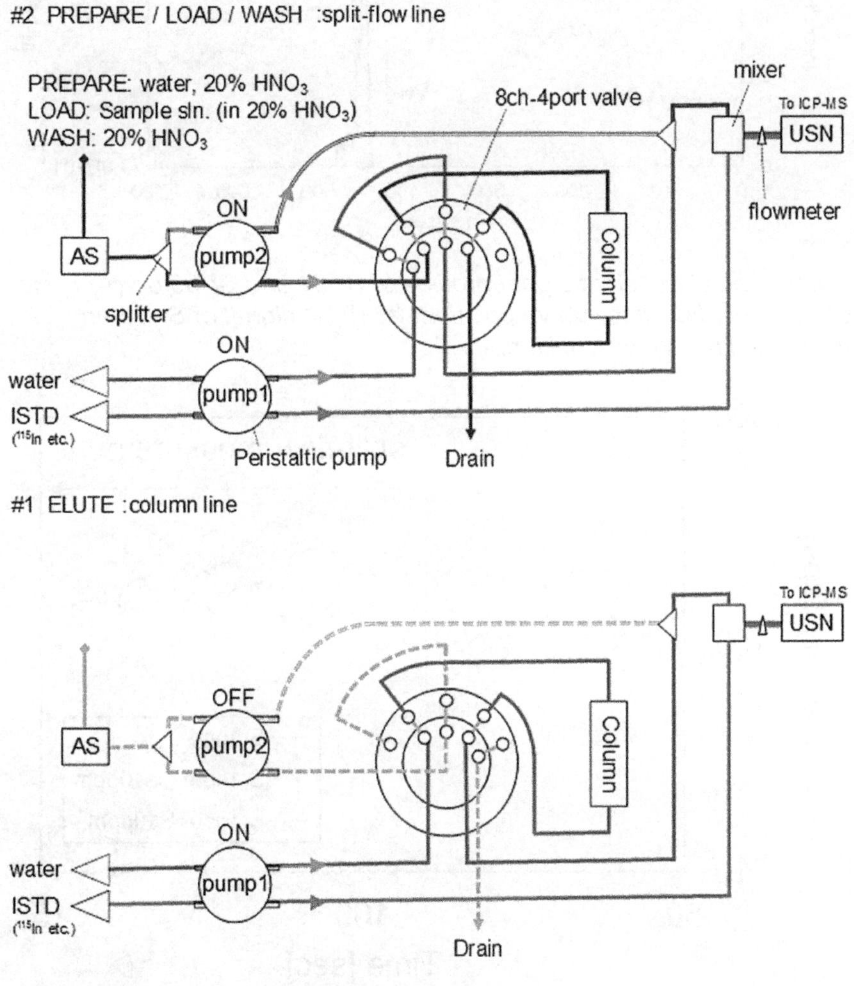

Figure 1 *Schematic of SPE-ICP-MS system.*

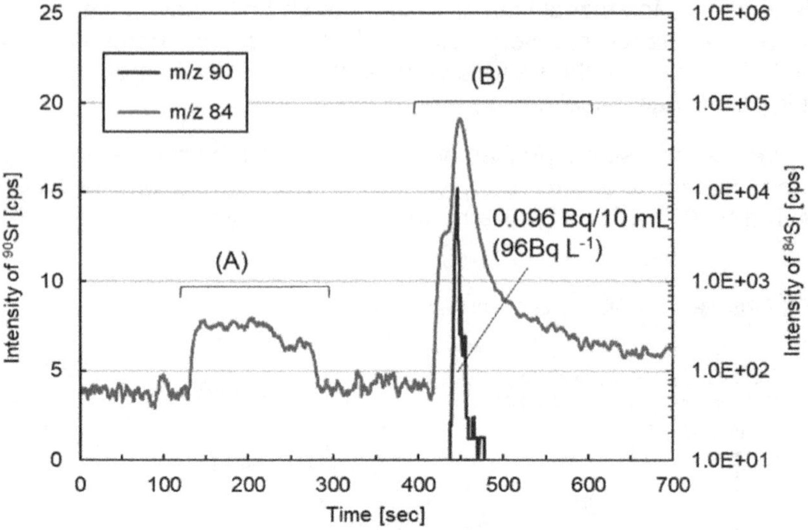

Figure 2 *Typical transient signal profiles of Sr-84 and Sr-90 on (A) is signal of Sr-84from split-flow line. Valve position: #2 (B) is signal of Sr-84 and Sr-90 from column line. Valve position: #1.*

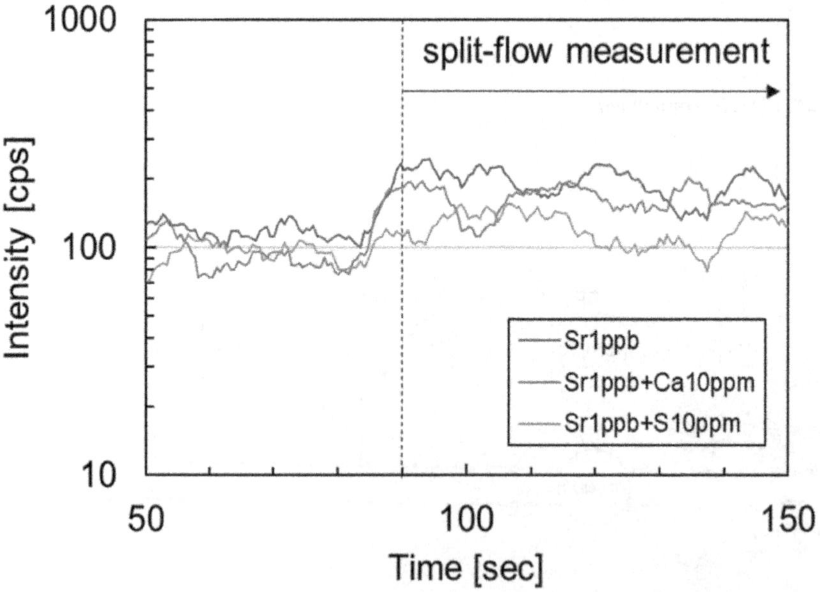

Figure 3 *Time dependence of intensity of Sr-84 from split-flow line.*

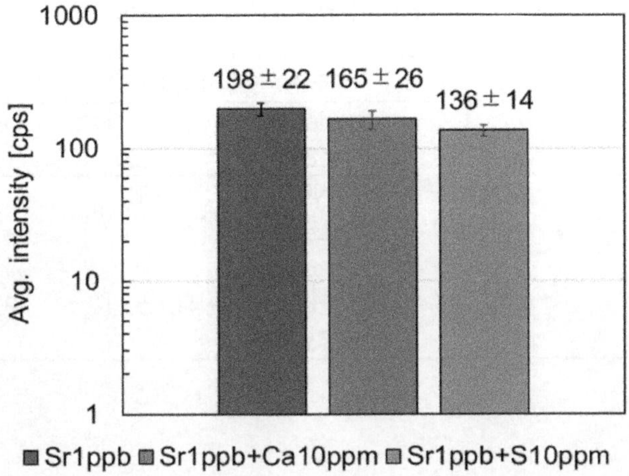

Figure 4 *Average signal intensity of Sr-84 from split-flow line.*

Figure 3 shows time dependence of the signal intensity of Sr-84 measured from split-flow line. Figure 4 shows the average of the signal intensity. There were no differences in signal intensity of Sr-84 between each samples. In the case of matrix concentration of these fish samples, it seems polyatomic species (e.g. ArCa, Ca_2 and SO_3) was no effect to Sr-84.

Table 2 shows the concentrations of Sr-90 in the solution digested fish and spiked Sr-90. Measured value showed good agreement with amounts of Sr-90 spiked in digested samples. R% has varied in the range of around 90% but did not seem to be particularly closely correlated with matrix concentration.

The detection limit of Sr-90 was calculated by using background of signal from column line and ratio of intensity to concentration in the standard solutions. In this system, detection limit (3σ) was 0.35Bq/L, and it is equivalent to 740 Bq/kg when Sr-90 in digested fish diluted 100 times is measured. In the case of bones, detection limit will be about 10 times higher because large amounts of Ca need to be diluted. In the case of solution, this detection limit is the applicability to WHO guidelines for drinking water. And in the case of fish, it shows the possibility of applicability to food monitoring (CODEX STAN 193-1995). For expanding adaptability such as monitoring at the time of accident, several improvements will be required to get lower detection limit and widen application range.

In the case of radioactivity measurement,[9] it takes 2 weeks or more (ion exchange: 2days or more, radiation equilibrium with Y-90: 2 weeks, liquid scintillation measurement: about 60 min) after sample is dissolved. On the other hand, in the case of ICP-MS measurement, it takes less than 1 day (ion exchange and ICP-MS measurement: about 60min).

Table 2 *Concentrations of Sr-90 in the solution digested fish.*

spiked [Bq L^{-1}]	measured [Bq L^{-1}], n=3			
	blank	digested fish, ×1000	digested fish, ×200	digested fish, ×100
0	ND	ND	ND	ND
12		8.0±4.6		
24		21.9±12.1		
48	48.0±6.1	48.1±12.1	44.7±20.2	39.4±11.0
96		108.5±11.1		

4 CONCLUSIONS

In order to confirm the applicability to food monitoring requiring rapidity, the concentration of Sr-90 spiked in fish was measured using the SPE-ICP-MS system. The detection limit (3σ) was evaluated as 0.35Bq/L and 740Bq/kg. liquid sample is applicable level to WHO guidelines for drinking (10Bq/L). For applicable level to food monitoring (100Bq/kg), several improvements will be required to get lower detection limit.

CONFLICTS OF INTEREST

There are no conflicts to declare.

ACKNOWLEDGEMENTS

The authors are grateful to Prof. Y. Takagai, and Dr. M. Furukawa for coaching of SPE-ICP-MS system.

REFERENCES

1. Y. Takagai, M. Furukawa, Y. Kameo and K. Suzuki, *Anal. Methods*, 2014, **6**, 355–362.
2. M. Furukawa, M. Matsueda and Y. Takagai, *Bunseki Kagaku*, 2017, 66, 181–187.
3. M. Furukawa and Y. Takagai, Anal. Chem., 2016, **88**, 9397–9402.
4. M. Furukawa and Y. Takagai, *Patent*, JP631530 B2, 2018.
5. J. Yoshinaga, M. Morita, M. Yukawa, K. Shiraishi and H. Kawamura, *J. AOAC Int.*, 2001, **84**, 1202–1208.
6. J. Irrgeher, T. Prohaska, R. E. Sturgeon, Z. Mester and L. Yang, *Anal. Methods*, 2013, **5**, 1687–1694.
7. N. Yamada and J. Takahashi, *Bunseki Kagaku*, 2018, **67**, 249–279.
8. T. W May and R. H. Wiedmeyer, *Atom. Spect.*, 1998, **19**, 150–155.
9. K. Fujimoto, *et al.*, *Environ. Sci. Tecnol.*, 2015, **49**, 7294–7301.

HOLDING TIMES AND PRESERVATION FOR ENVIRONMENTAL RADIOCHEMICAL SAMPLES: AN EVALUATION OF ISO STANDARD GUIDELINES

Jasper Hattink and Roger Benzing

SOCOTEC UK Ltd, Unit 12, Moorbrook Industrial Estate, Didcot, Oxfordshire OX11 7HP, UK.

* Jasper.Hattink@socotec.com

ABSTRACT

The condition of a sample taken and stored before analysis is an important aspect in analytical measurements including in the field of environmental radiochemistry. The validity of the result from the test depends heavily on the sample integrity; a sample that is not well preserved might not be a true representative of what has been sampled. For example, the radionuclides in a water sample might be adsorbed by the wall of the container during transport and storage. This study summarises a range of shelf-life tests performed by SOCOTEC nuclear laboratory. Our studies suggest that the integrity of freshwater samples can be preserved for up to 1½ year when filtered and acidified for the analyses of gamma emitting radionuclides and at least 6 months for actinides. Tritium, either in its natural non-acidified state or acidified, can be stored for at least 3 years. For a minimum loss of the transuranium actinides, filtration and acidification should occur within 3 to 7 days of sampling. Cooling the samples during transport and interim storage is advised to suppress bacterial and algal growth.

1 INTRODUCTION

One of the important aspects wen performing a (radio-) analytical measurement is the condition of a sample taken and stored. Samples not well preserved may not be a true representative of what has been sampled if storage conditions affect the integrity of the samples. Such might, in turn, question the validity of a measurement. For example, the radionuclides in a water sample may be adsorbed by the wall of the container during transport and storage.

This important aspect was also recognised by the Laboratory Committee of the European co-operation for Accreditation (document EA/LC(07)60)[1]. Laboratories working under ISO 17025 or ISO 15189 are recommended to apply the best available sample storage techniques. ISO 5667-3:2018 gives a full overview of the best available techniques for a range of chemical parameters in water samples and summarises both validated and tested storage and handling techniques including maximum holding, or, shelf-life times[2]. There are however some gaps in the recommendations for storage and preservation of water samples for radiochemical analyses.

This study summarises a range of shelf-life tests performed by the SOCOTEC UK nuclear laboratory. Each test evaluated the integrity of a group of radionuclides in water samples under different storage conditions. Samples were aqueous certified reference materials (CRM), proficiency test samples or spiked freshwater samples, depending on the availability.

2 EXPERIMENTAL

2.1 Overview

The experimental work encompasses three different data sets. All work was performed on freshwater, either simulated (Tritium tests), sampled from a stream (Gamma spectrometry test) or tap water (Actinide tests). Samples were analysed by established methods, either published as ISO standard or developed in-house based on published methods in open literature under the overarching quality standard ISO 17025.

2.2 Tritium shelf-life tests

Both proficiency test samples and certified reference materials were available for this test. A series of proficiency test samples and certified reference material were analysed within 2 weeks of receipt and after set storage periods (after 44, 99, 107, 160, 204, 295, 350, 421, 477, 575, 743, 939, 1078, 1200, and 1261 days). Proficiency test samples were provided by LGC, Teddington within the AQUACHECK scheme between 2012 and 2015. These samples were provided in glass bottles, and after analysing half of the sample, stored in a fridge at 4°C. The certified reference material used was seawater IAEA-443 and provided by the IAEA, Vienna in a 5 L plastic container and stored at room temperature. It arrived in 2015 and was kept at room temperature for a year. Samples were analysed according to ISO-9698 Annex D,[3] after a total storage time of 3469, 3483, 3500, 3514, 3531, and 3545 days. A 100 mL test portion was distilled under alkaline conditions using sodium carbonate to suppress the formation of iron hydroxide. The first 30 to 35 mL was discarded, and the subsequent 30 mL collected for analysis.

Ten to 12 mL was dispensed into scintillation cocktail. Over the space of the trial period, either Instagel from Perkin Elmer, Goldstar multipurpose scintillation cocktail, Meridian, or Ultima Gold LLT, Perkin Elmer was used.† In some cases only a limited sample amount was available. In those cases, only a 25 mL test portion was used and the entire sample distilled, as described in ISO-9698 Annex E[3].

2.3 Gamma ray emitting radionuclides shelf-life tests ("Gamma ray spectrometry")

The shelf-life for the gamma ray emitting radionuclides ^{60}Co, ^{137}Cs, ^{226}Ra and ^{241}Am was tested in freshwater. Freshwater was sampled from a small brook in Oxfordshire, UK (Moor Ditch at 51°37′01.0"N, 1°14′50.5"W). Water samples were filtered using a 0.45 μm filter (Sarstedt) to remove suspended solids and stored in 2 L white high density polyethylene containers. The pH of this water was 7.7±0.1. Samples were spiked with the respective radiotracers at a nominal concentration of 1 Bq L^{-1}. One set was stored with no preservative added and a second set was acidified with nitric acid to a pH of 2.7±0.1. Samples were stored for up to 365 days. During this period, no change of pH was observed.

Samples were analysed within 15 minutes of preparation and after a set storage time of 1, 2, 3, 4, 5, 7, 10, 14, 21, 28, 60, 100, 189, 365, and 490 days using high resolution gamma ray spectrometry. At each storage interval, duplicate samples of 500 mL each were placed in Marinelli beakers and measured for a 7 hour measurement period. FitzPeaks gamma ray spectrometry software was used to analyse the gamma ray spectra obtained.

2.4 Actinide shelf-life tests

The shelf-life for the actinides ^{234}U, ^{235}U, ^{238}U, ^{239}Pu, ^{241}Am and ^{244}Cm was tested using tap water. Samples were stored in 1 L transparent high density polyethylene containers. One set was stored with no preservative added (pH of 7.5±0.1) and a second set was acidified with nitric acid to a pH of 2.0±0.1. Samples were analysed within 15 minutes of preparation and after a set storage time of 26, 28, 52, 63, 66, 75, 73, 80, 84, 88, 105, 110, 116, 125, 143, 147, 160, 171, and 192 days.

Samples were using alpha spectrometry following chromatographic separation of the actinides. Yield tracers for uranium (^{232}U), plutonium (^{242}Pu) and americium and curium (^{243}Am) were added to correct for analytical losses. After an initial co-precipitation on iron hydroxide at pH of 5, the actinides were separated using a strong anion exchange resin (DOWEX X-8, Dow Company) on a nitric acid gradient. The separated uranium fraction as well as the americium and curium fractions underwent a second polishing using a nitric

---

† All scintillation cocktails used had a satisfactory performance, but the high loading capacity of the Ultima Gold LLT combined with its greater stability has recently been preferred over the other two.

acid gradient on either DOWEX X-8 for uranium or TRU-resin (TrisKem) for americium and curium. Actinides were electroplated onto stainless steel disks using a sulphate buffer at 1000 mV. Electroplated counting sources were placed 1 cm from the detector and measured for 250,000 seconds to collect the alpha spectra. Alpha spectrometry systems using PIPS detectors or SSB detectors were used to collect the alpha spectra. Typical detector efficiencies were 15% to 17% and chemical yields varied between 50% and 70%. Spectra were analysed using set regions of interest and activity concentrations either calculated manually using Excel (Microsoft) or automated using Alpha APEX (Canberra) or Alpha Vision (Ortec).

3 RESULTS AND DISCUSSION

3.1 Tritium shelf-life tests

Our studies show that tritium was stable for a much longer period of storage than that recommended in ISO 5667-3:2018[2]. Tritium was generally stable for up to 3 years. We showed that in seawater, the tritium concentration was stable for at least 6 years. Tritium concentration was not affected by different storage temperature (ambient or cold), container type (glass or plastic), acidification, or the presence of head space.

Figure 1 shows the effect of storage on the tritium activity concentration. An increase in tritium concentration after 1000 days is suggested, but similar results were recorded at 100 and 400 days. The correlation between the data measured on receipt and after storage is high; i.e. 0.998, indicating no systematic effect of storage for a prolonged time.

Acidification of the samples didn't influence the tritium activity concentration, as was demonstrated by a comparison between the CRM seawater and the PT scheme freshwater (T-test, α=0.05, P=0.25) and confirmed in a separate test on one sample, of which 6 test portions were acidified and 6 test portions were un-acidified (T-test, α=0.05, P=0.1, data not shown). This confirms the advice given in 'Technical Notes for EPA Method 906.0 Tritium in Drinking Water'[4]. When samples are preserved with acid to a pH of 2, the amount of hydrogen added is so small relative to the total amount of hydrogen present that it does not significantly change its concentration in the sample by more than about 0.2%. Additionally acid present in the concentrations used for sample preservation will be neutralized during the tritium analytical procedure.

3.2 Gamma emitting radionuclides shelf-life tests ("Gamma ray spectrometry")

The gamma emitting radionuclides ^{60}Co, ^{137}Cs, ^{226}Ra, and ^{241}Am were stable in acidified filtered freshwater (pH 2 to 3) for up to 1½ years in plastic containers.

Figure 2 summarises the effect of storage for gamma emitting radionuclides of ^{60}Co, ^{137}Cs, ^{226}Ra and ^{241}Am in either non-acidified water samples (pH of 7.7±0.1) or acidified samples with nitric acid (pH of 2.7±0.1). The effect of acidification as preservation can be clearly seen; the activity concentration for all radionuclides are stable for at least up to 1 year. In the non-acidified sample, a rapid decrease of ^{241}Am can be seen. Within 10 days of storage under non-acidified conditions, the ^{241}Am concentration drops by 40%, most likely due to adsorption to the container wall. After 10 days, an equilibrium between the container wall and the water sample seems to be established. ^{60}Co stays for the first 14 days of storage in non-acidified conditions. Hereafter, a slow decrease in activity concentration can be observed, most likely due to microbiological growth in the container. ^{137}Cs and ^{226}Ra seems to be stable under both acidified and non-acidified conditions for at least 1 year. The variability in the ^{226}Ra activity can be fully explained by the measurement uncertainty.

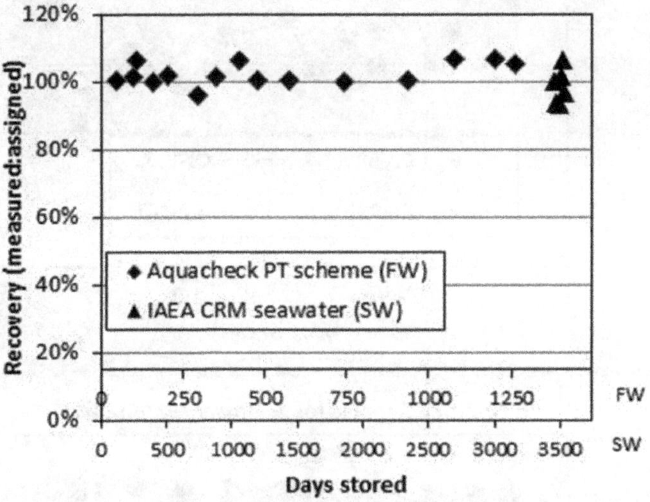

Figure 1 *Recovery of tritium after storage relative to the assigned value. Uncertainty of measurement is 3% (1 sd). Note that the freshwater samples were stored at 4°C in glass with headspace present for up to 1250 days (top time axis indicated with "FW"). The Seawater CRM was stored for about 3500 days and was acidified and kept at room temperature in a 5 L plastic container (bottom time axis indicated with "SW").*

3.3 Actinide shelf-life tests

The actinides uranium, plutonium, americium, and curium were stable for at least 6 months in filtered and acidified freshwater (pH of 2) in plastic containers. This is longer than the 2 months recommended in ISO 5667-3:2018[2].

Figure 3 shows the effect of preservation with nitric acid for selected alpha emitting radionuclides; ^{234}U, ^{235}U, ^{238}U, ^{239}Pu, ^{241}Am and ^{244}Cm. A very similar behaviour can be seen for plutonium, americium and curium; all these actinides show a rapid decrease of activity concentration. The decrease was not as fast as observed in the white 2 L container (see Figure 2), but is still significant, i.e. after 80 days, a decrease of 8%, 7% and 24% for ^{241}Am, ^{244}Cm, and ^{239}Pu, respectively, could be observed, compared to 30% decrease for ^{241}Am in the white container (measured by gamma spectrometry). The reason for this difference might be found in the type of container (a white container versus a clear container) and container surface. The white container that was used for the gamma spectrometry tests has a surface of about 960 cm², whilst the surface of the round 1 L bottle is about 500 cm².

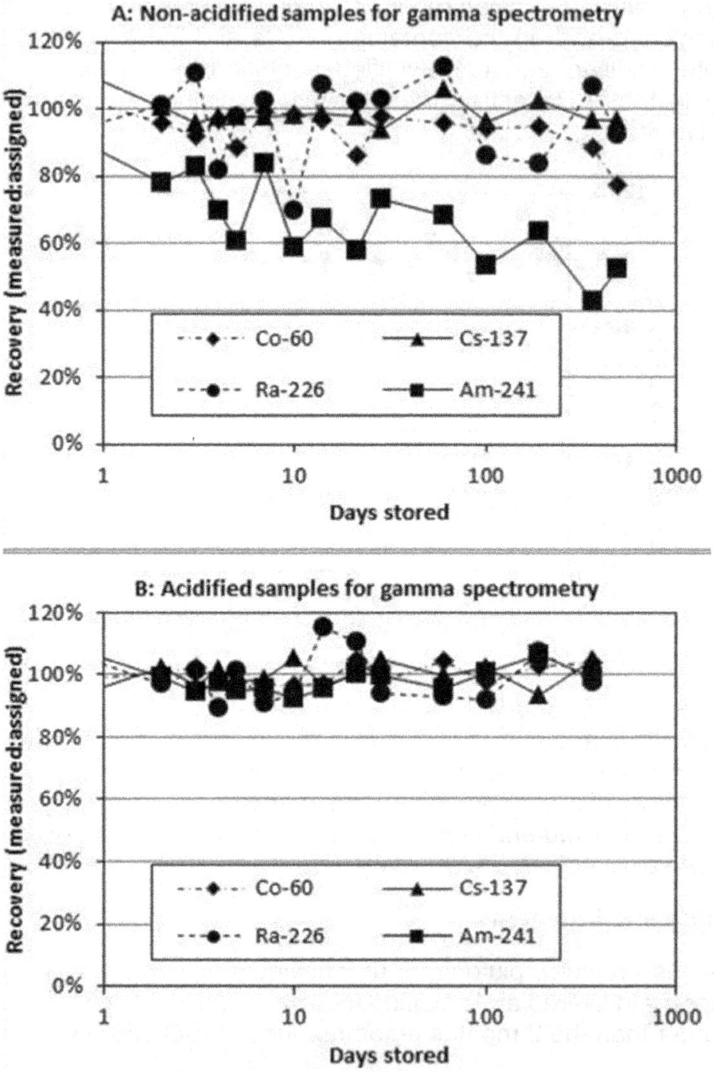

Figure 2 . *Stability of selected gamma emitting radionuclides;* ^{60}*Co,* ^{137}*Cs,* ^{226}*Ra and* ^{241}*Am. Panel A shows the effect of storage on the activity concentration in non-acidified samples (pH=7.5), Panel B shows the preserving effect of the acid added (final pH=2.7) on the storage of these samples. Final data point was taken after 490 and 365 days of storage, respectively for non-acidified and acidified samples.*

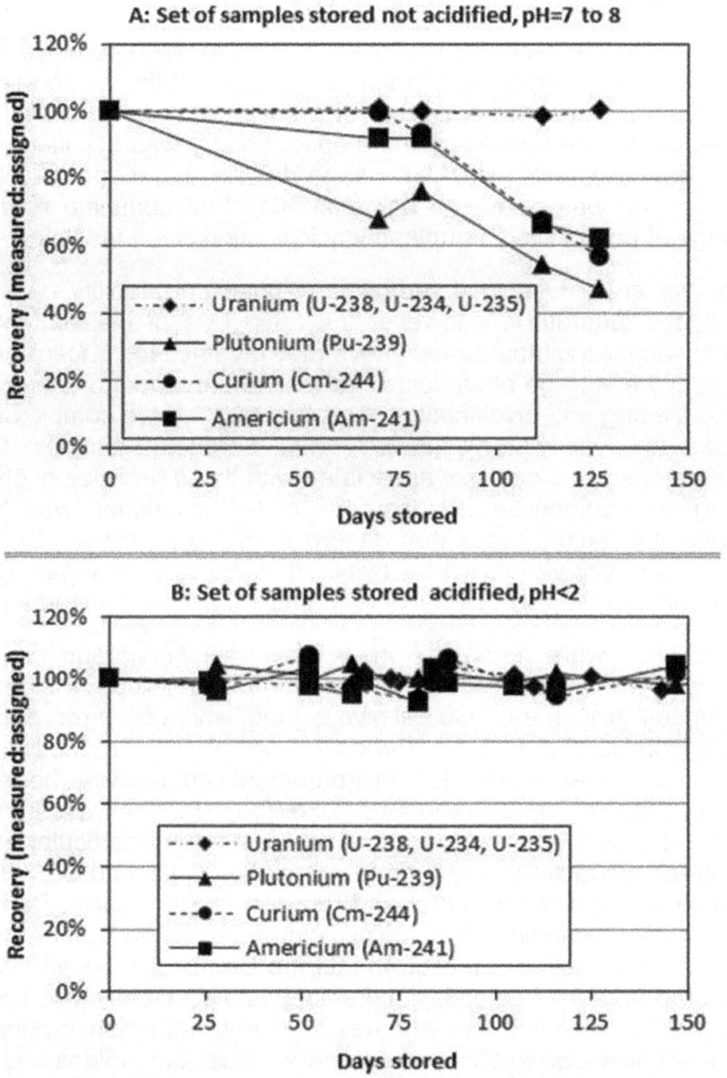

Figure 3 *Recovery of selected alpha emitting radionuclides; ^{234}U, ^{235}U, ^{238}U, ^{239}Pu, ^{241}Am and ^{244}Cm. Panel A shows the effect of storage on the activity concentration in non-acidified samples (pH=7 to 8), Panel B shows the preserving effect of the acid added (final pH<2) on the storage of these samples.*

Secondly, the colorant or pigment used could increase the adsorption sites. It is unknown to the authors which colorant has been used, but if, for example an inorganic pigment like TiO_2 is used, it could create additional adsorption sites on its oxygen groups, similar to the complex formation reactions at natural solid-solution interfaces[5].

3.4 General comments and considerations

If adsorption and desorption processes are responsible for the majority of losses, it could be possible that through acidification of the whole sample in its container, the radionuclides could be recovered. At the end of the 490 days storage period, the sample for gamma spectrometry was acidified and left to leach for 1 day. Some of the ^{241}Am was re-dissolved, but about 20% was still missing from the water sample, demonstrating that complete re-dissolution after storage of un-acidified samples by acidification is not feasible.

The above study was to determine the chemical stability on storage of samples in the laboratory. However the rapid loss of the lanthanides and actinides in samples without preservative may give rise to concern of potential losses during the storage of samples in transit to the laboratory after sampling but prior to filtering and acidification. This is a much more complex issue due to samples of river water, seawater and borehole samples containing suspended solids. Radioactivity associated with these particles might act as a buffer and all radionuclides adsorbed to the container wall might be "replenished" by radionuclides that desorb from the particles. On the other hand, if particles show biological activity, as in the case of algae, desorption might be counteracted by bacterial growth and subsequent adsorption.

Radionuclides were added to show the effect of storage on individual radionuclides, but can be also act as a surrogate to groups of elements with similar chemical properties. ^{60}Co will give a good, worse case representation of the metals from Group 7 to 12, Period 4 and for Al, Cd, Sn and Au. The elements of particular interest for environmental radioactivity measurements are Mn, Fe, Co, Ni, Zn, Cd, and Sn. Likewise, ^{137}Cs represents the Alkali metals, Group 1 of the periodic table. The elements of particular interest for environmental radioactivity measurements are Na, K, Rb and Cs. ^{226}Ra is the least soluble element of the group and represents the Alkaline Earth metals, Group 2 of the periodic table. The elements of particular interest for environmental radioactivity measurements are Be, Ca, Sr, Ba and Ra. ^{241}Am represents the Rare Earth metals; lanthanides and actinides make up the inner transition metals. The lanthanide series, elements with atomic number 57 to 71 and the actinides, 89 to 103 also Sc and Y. Most lanthanides and actinides are present in a 3+ oxidation state and will behave chemically in a very similar way. Some lanthanides and actinides display a redox chemistry; Ce, U, Pu, with the highest oxidation state more soluble than the lowest oxidation state. The elements of particular interest for environmental radioactivity measurements are Y, Ce, Pm, Sm, Eu, Ac, Th, U, Np, Pu, Am and Cm.

4 CONCLUSIONS

We found that acidification is critical for good preservation of most of the gamma emitting radionuclides and the actinides. If the water is not acidified, ^{241}Am concentration decreased rapidly within a week. After one month of storage the ^{241}Am concentration decreased by 60% to 65%. A similar result was seen for plutonium and curium. However, ^{60}Co concentration decreased at a much slower rate over time; after 1½ years 5% to 10% was lost. Loss of radionuclides could not be recovered by acidifying the container afterwards. For some radionuclides, acidification is not as critical. For example, the uranium concentration was stable for at least 4 months in the non-acidified samples, whilst ^{137}Cs and ^{226}Ra were stable over a 1½ year sample storage period.

Our studies suggest that the integrity of freshwater samples for radiochemical analyses can be preserved for up to 1 year when filtered and acidified for the analyses of gamma ray emitting radionuclides, actinides and tritium. If filtration and acidification cannot be achieved in the field, for logistical and health and safety reasons, samples should be transported as soon as possible to a suitable laboratory for filtration and acidification. For a minimum loss of the transuranium actinides, this should occur within 3 to 7 days. This period can be extended if these are not of interest. Cooling the samples during transport and interim storage is advised to prevent bacterial and algal growth as much as possible, to avoid an increase in particulates and a subsequent absorption, including possible biological uptake.

CONFLICTS OF INTEREST

There are no conflicts to declare.

ACKNOWLEGEMENTS

The authors wish to thank Claire Foulon, Charlene Hunston, Ruhina Miller, Lauren Perrie, and Patrick Williams for support with the sample preparation and analyses.

REFERNCES

1. United Kingdom Accreditation Service, *TPS 63. UKAS policy on deviating samples*, UKAS, Feltham, 2013.
2. International Organisation for Standardisation, *ISO 5667-3:2018 Water Quality – Sampling – part 3: Preservation and handling of water samples*, ISO, Geneva, 2018.
3. International Organisation for Standardisation, *ISO 9698:2015 Water Quality – Determination of tritium activity concentration – Liquid scintillation counting method*, ISO, Geneva, 2015.
4. United States Environmental Protection Agency, *Technical Notes for EPA Method 906.0, Tritium in Drinking Water,* US EPA, Washington, 2009
5. W. Stumm, J.J. Morgan, *Aquatic chemistry. An introduction emphasizing chemical equilibria in natural waters*, John Wiley & Sons, New York, 1981

APPLICATION OF URANIUM RADIO-CHRONOMETRY TO INTERPRET URANIUM SAMPLES OF KNOWN PROVENANCE

Matthew Higginson,* Paul Thompson, Brogen Dawkins, Fiona Taylor and Phillip Kaye

AWE, Aldermaston,RG7 4PR

* Matthew.Higginson@awe.co.uk

ABSTRACT

Radio-chronometry is an important signature in understanding the processing history of an unknown nuclear material. Knowledge of the formation of processing signatures is crucial to underpin nuclear forensic analysis. A precise concordant age from multiple chronometers gives confidence in the model age and can aid in determining processing history. Discordant chronometers can also help to refine specific processing histories where incomplete processing is known to occur. In this work we apply our rapid radiochemical methodology to materials of known provenance to further validate our capability. In these analyses we aim to study the effect of known processing history on model ages to build confidence in our understanding of the signature. We observe concordant chronometer pairs and agreement with processing history in samples where the model age assumptions are valid, such as purification by solvent extraction and specific processing techniques. This gives confidence in the utility of the analysis for forensic applications. Where discordance was observed in cast uranium metal samples, the signature was valuable for interpreting the processing history in terms of thorium and protactinium.

1 INTRODUCTION

Nuclear forensics is the analysis of intercepted nuclear, radiological, and other materials to support legal cases relating to nuclear security. The field has been highlighted as a research focus area by the international nuclear security community.[1] As part of an overall security strategy the methodology serves to strengthen safeguards and to deter illegal trafficking. A justification for this is that it can allow for the determination of the origin of an unknown material.[2]

The processing and illicit use of nuclear materials imparts specific indicators in the chemical composition of a sample. These relationships can be used to interpret the history and intended use of the material.[3] Case studies and reviews focusing on the methodology, casework, and techniques have been published.[4,5] Historically, uranium samples are the most commonly intercepted and analysed, therefore the analysis of signatures in these materials constitutes an area on which to focus fundamental research.[6]

Flexible, robust and rapid methodologies are advantageous to ensure timely, high quality, defensible results for nuclear forensics. Flexible techniques are also valuable as they can be readily adapted and validated to the complex sample types which may be encountered.[7]

Nuclear forensic analysis relies on understanding the evolution of signatures during processing. A technique to build confidence in the interpretation of signatures in an unknown material is the analysis of materials of known provenance, and processing history.[8]

Radio-chronometry is the quantification and modelling of decay progeny in a sample. It is a key signature addressing the "when" in nuclear forensics. The model age calculated is based on many assumptions.[9] These have been reported in detailed studies.[10,11] The key assumption in the calculated age is that decay progeny are completely removed at the time of production and the sample has behaved as a closed system.

Multiple chronometer pairs are analysed by forensic laboratories to build confidence that the model age represents the true time of last processing. If agreement (concordance) or disagreement (discordance) is determined, this aids in the understanding of chemical fractionation during the nuclear fuel cycle. In the analysis of uranium, the ^{238}U and ^{235}U decay chains are typically investigated using the $^{231}Pa/^{235}U$ and $^{230}Th/^{234}U$ decay pairs. This is because they are amenable to accurate determinations from a range of material types and enrichments.

Precision analysis is essential for radio-chronometry as it allows discrimination between samples of similar ages and can confine the age of the material to aid in an investigation.

We achieve this measurement using isotope dilution (tracers of ^{229}Th, ^{233}U), selective separation, certified reference materials (U630 CRM) and sensitive instrumental methods. The measurement used depends on the reporting timescales and requirement of the customer. We use isotope dilution alpha spectrometry (ID-AS), thermal ionisation mass spectrometry (TIMS) and multi-collector inductively coupled plasma mass spectrometry (MC-ICP-MS).

In a previous report, we demonstrated the development of a novel rapid radiochemical method for the analysis of the $^{231}Pa/^{235}U$ and $^{230}Th/^{234}U$ decay pairs from a single sample. The method uses an initial stacked column of TK400-AGI-X8-UTEVA to purify fractions of Pa, U and Th respectively, allowing for additional purification for multiple detection systems and turnaround timescales. The system uses an in-house control material, certified reference materials, and vacuum assisted separations.[12]

This paper aims to highlight samples analysed by our recently reported age dating method with the aim of understanding the apparent model ages with respect to known production histories and concordant and discordant chronometer pairs.

2 EXPERIMENTAL

The separation scheme is reported in detail in a publication on method development.[12] Expanded uncertainties were reported at the k=2 confidence level unless otherwise stated. The U630 CRM was analysed with each batch and was shown to be within the certified value. Analysis was completed by alpha spectrometry for $^{229/230}Th$ and ^{231}Pa. An Ortec alpha ensemble, with Alphavision 7.0 software, calibrated using certified mixed uranium/plutonium standards was used for analysis. ^{233}Pa was determined by gamma spectroscopy using an Ortec high purity germanium gamma spectrometer calibrated with certified mixed gamma standards. Determination of uranium isotope ratios was achieved by TIMS by loading 10 ng of uranium sample onto a zone refined, carburised rhenium single filament and running an automatic peak jumping method on a Thermo Triton to collect data. NBS U005a was used for mass-bias correction. The assay of the control materials and certified reference materials was carried out using a modified Davis and Grey method on a Metrohm auto titration system alongside certified assay standards to give weight percent uranium to 0.10% standard error.[13] All chemicals used for radio-chronometry were of Optima grade (Fisher). Chemical yield tracers and reference materials were provided from calibrated stocks within AWE. Uranium (U_3O_8) radio-chronometry and isotopic standard reference materials U005a, U630 and U930 from New Brunswick Laboratory (NBL) were used in this study. Tracers SRM 4342 (^{230}Th), SRM 4328c (^{229}Th), Isotope Products Laboratories (^{233}U) Amersham UK NGZ44 (^{237}Np) were used for isotope dilution, yield tracers or for purification of ^{233}Pa. The radionuclide concentrations were quantified by ID-TIMS and ID-AS. Ultrapure perfluoro alkoxy alkanes (PFA) bottles (Romil) were used for all sample transfers and elution from columns. Extraction chromatography resins were purchased from TrisKem and washed with UHP reagents prior to use. All samples were taken in duplicate, spiked and un-spiked, with a combined process blank used.

3 RESULTS AND DISCUSSION

The model age calculated from a radio-chronometry measurement will ideally describe the production, purification or processing age of the material. To continually validate our separation methodology on different material types, we analysed UO_2, U_3O_8, uranium metal and materials of varying enrichment.

3.1 Depleted uranium metal purified by solvent extraction

Two batches of depleted uranium metal samples were dissolved and analysed by our previously reported methodology.[12] This sample was known to have been purified by solvent extraction after reprocessing and reduced to metal by casting using vacuum induction melting. We predicted this processing would effectively remove thorium and protactinium from bulk uranium and that casting would also introduce an additional separation stage in the material. The results are shown below (Figure 1).

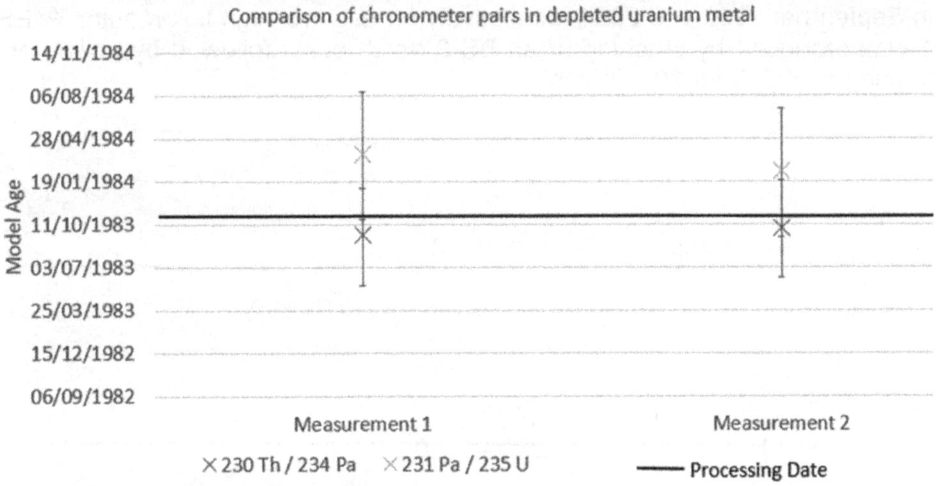

Figure 1 *Model age of depleted uranium metal sample in this study.*

The measurement of Pa in these samples required long counting times by radiometric methods due to the low abundance of ^{235}U (<0.4%). MC-ICP-MS measurements for protactinium are advantageous for a quicker turnaround and higher precision analysis for this type of sample. The uncertainties on the $^{231}Pa/^{235}U$ chronometer are also larger due to the limited availability of certified reference materials and calibration of a ^{233}Pa spike. Planned measurements by MC-ICP-MS may also allow us to investigate the older trend for the $^{231}Pa/^{235}U$ chronometer in this material.

The uncertainties overlap to within +/- 4 months by this measurement and both chronometers are concordant. This supports our hypothesis that this material was almost completely purified from daughter products during production. This also aids us in determining that this solvent extraction purification will give a concordant signature in an unknown sample. The material was reprocessed and cast in 1983 into flat pieces of metal and the two measurements were taken from separate pieces. This also aids in understanding that this process produced a homogenous material with respect to Th and Pa segregation.

3.2 Example of concordance in variable enrichment fuel cycle samples

We have also observed agreement between the model age and production history in studies with fuel pellets produced from high-fired oxides. The international collaborative material exchanges (CMX) for the international working group on nuclear forensics (ITWG-NF) have implemented two sets of fuel pellets in separate exercises.[14]

In the CMX-4 exercise, the samples provided were one powder (ES-1) and two pellets (ES-2 and ES-3) from two batches of enriched material (2.2 and 2.9%). The pellets were produced from UF_6 by AREVA in two separate batches in September 2002 and 2004 respectively for use in a high flux reactor.[15] ES-1 was produced by crushing of an ES-3 batch pellet followed by milling and heating at 325°C for 90 minutes.

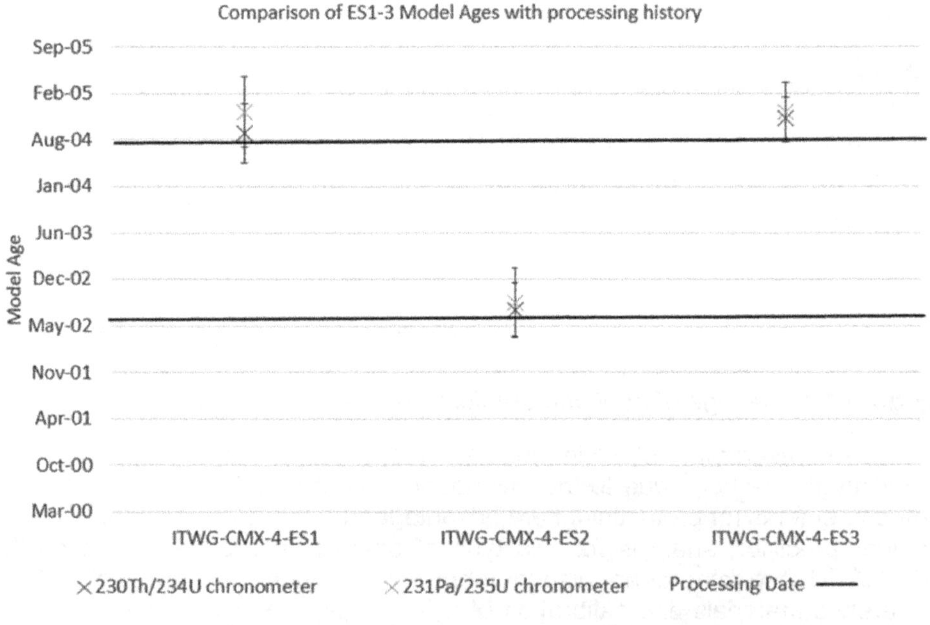

Figure 2 *Comparison of ES1-3 samples in this study.*

In this case the chronometer pairs are concordant in all samples and the precision of the analysis allows for the distinguishing of two distinct sets of samples and processing dates. The model ages therefore agree with the processing history of a highly processed fuel cycle material. The processing from UF_6 and subsequent heat treatment into the associated oxide purified the majority of thorium and protactinium from bulk uranium in the material.

The results again yield concordant chronometer pairs, giving confidence that the ages agree with the date of last production and that our method produces quality results for interpretation.

3.3 Discordance in early fuel cycle materials

The results for early fuel cycle materials are less reliable for any pairs based on complete purification techniques (i.e. $^{230}Th/^{234}U$ and $^{231}Pa/^{235}U$). This is due to residual decay products in the material leading to bias towards older ages than the known production date. An innovative approach based upon the $^{232}Th/^{228}Th$ decay pair in samples was reported by Varga *et al.* for uranium ore concentrates (UOCs) and was shown to be consistent for samples less than approximately thirty years old.[16] This is due to insensitivity to residual decay progeny by Pa, U and Th. This approach was shown to give accurate production dates for samples less than thirty years old and is only limit of quantification and quantity of ingrown ^{228}Th in the sample. In the case of older samples, as they approach secular equilibrium the curve can only be solved for the minimum age of the material.

UOC samples from the early fuel cycle has shown both trends. Two data sets from UOCs in our library were evaluated in this study. (Table 1) The model ages do not agree with the known operation dates of the source mines.[18] In addition the results for Madawaska do not agree with other reported ages in the literature.[17] We believe this to be from inhomogeneity in sampling of different batches of materials and the smaller sample sizes (0.1 g) employed in our study. In this case, radio-chronometry can be used to limit the minimum age of these types of material.

Table 1 *Radio-chronometry data from two UOC samples by the $^{230}Th/^{234}U$ and $^{228}Th/^{232}Th$ chronometer.*

UOC ID	Reference Date (09:00:00 GMT)	Processing Date $^{230}Th/^{234}U$ chronometer (Years k=2)	Uncertainty (Years k=2)	Estimated Processing Date $^{228}Th/^{232}Th$ chronometer	Mine Operation Date
Canada Madawaska	01/08/2018	Apr-67	2.13	Before 1988	1975–1982
Canada Faraday	01/08/2018	Nov-54	2.31	Before 1979	1954–1964

The results show that apparently older model ages are consistent with the types of processing used in the early fuel cycle where it is not essential to remove Pa or U for product specification. This information, along with additional forensic signatures related to UOCs, such as rare earth element pattern, geolocation, and trace elemental impurities would allow comparison with data on known operating periods of uranium mines.

3.4 Studies of cast metal sample

Two batches of highly enriched uranium metal samples were also analysed as part of this work, the material was known to have been cast in the early 1990s from varied feed materials. The sample was taken in duplicate in two separate batches of analyses (Figure 3).

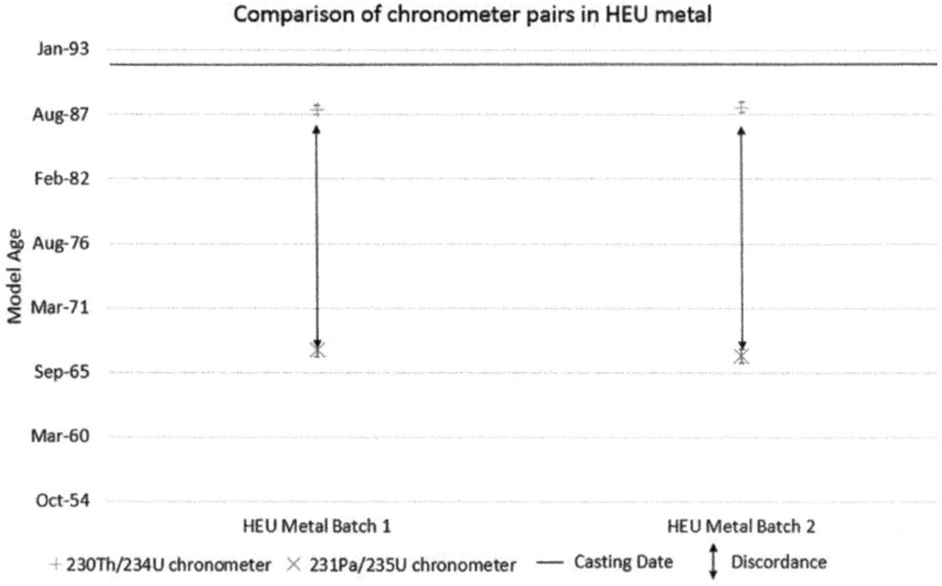

Figure 3 *Comparison of chronometer pairs in two HEU metal samples, example of chronometer discordance.*

The method yielded the certified value for the CRM in both studies. The $^{230}Th/^{234}U$ model purification dates for the sample were concordant between the batches. This would suggest in an unknown sample that metal purification which purified thorium took place in late 1987. It is also plausible that incomplete separation of thorium from bulk uranium in part of the batch of feed material lead to an older apparent age based on material provenance.

Studies have predicted that thorium is separated during casting because of the density and metallurgical differences between thorium and uranium.[18] It is likely therefore that the thorium purified in the material was discarded as production waste. We plan in future studies to cast uranium under controlled conditions to investigate how casting influences the minor actinides.

The $^{231}Pa/^{235}U$ model purification dates between batches agree with a 20–25-year discordance with respect to the thorium chronometer and the known casting date. This implies that the metal casting procedure used with the feed material did not efficiently separate Pa during casting. As we sampled from approximately 0.5 g in each case we appear to have a homogenous Pa fraction. This discordant trend has also been observed previously in the ITWG CMX-3 metal pin sample from a fuel log cast at Y-12.[19] An additional finding of those studies by Kayzar *et al* demonstrated that between batches the $^{231}Pa/^{235}U$ chronometers were discordant.[20] This suggested that the processing method can lead to varying protactinium heterogeneity in the material. Additional smaller scale studies of this material are planned to investigate whether spatial heterogeneity in cast material can leave this signature.

Based on this study the results suggest that the uranium feed material was last purified in the late 1960s. As we do not know the complete history of the feed for this material it is not possible to verify this conclusively. If the material is blended this can also lead to a composite signature from the combination of additional decay progeny in the feed materials. The method and signature are still useful in this case as apparent discordance in an unknown material could be consistent with this type of processing and could be verified by analysis of trace elemental impurities.

4 CONCLUSIONS

Our rapid age dating methodology for the $^{231}Pa/^{235}U$ and $^{230}Th/^{234}U$ chronometers have been applied to additional sample types and have been shown to produce model ages that agree with certificate values and sample production history. We plan to continue to test and develop our method on samples provided to our laboratory. Additional chronometer pairs in the uranium decay series would be useful to add to the age dating tool kit.

The model age via the two decay pairs has been shown to accurately predict the known production history in processed materials such as reference materials, high fired oxides and fuel pellets. This adds to the utility of this method as a nuclear forensic signature.

Certain sample types require additional analysis times or large samples, for example Pa in 'young' depleted uranium for radiometric analysis. Uranium metal samples were shown to be discordant with known production histories. This was due to incomplete purification of both Th and Pa in the casting process. In these cases, care must be taken to interpret chronometric measurements. The magnitude of discordance between chronometer pairs can however be linked to the production history. Further studies of casting signatures for uranium would be valuable to aid in the interpretation of unknown uranium metal samples.

CONFLICTS OF INTEREST

There are no conflicts to declare.

ACKNOWLEDGEMENTS

The research herein was funded and supported by AWE plc, UK. James Dunne, Chris Gilligan and Nathan Thomas are thanked for help guiding this work.

NOTES AND REFERENCES

1. http://www.nss2016.org/2016-action-plans, accessed 22/10/2018.
2. G. Allison, MIT Technology Review, 2008, pp. 68–73.
3. M. Kristo, A. Gaffney, N. Marks, K. Knight, W. Cassata and I. Hutcheon, Annual Review of Earth and Planetary Sciences, 2016, 44(1), 555–579.
4. K. Mayer, M. Wallenius and I Ray, Analyst, 2005, 130, 433–441.
5. K. Mayer, M. Wallenius and Z. Varga, Chem. Rev., 2012, 113, 884–900.
6. IAEA Incident and Trafficking Database (ITDB) Factsheet 2017. Available at: https://www.iaea.org/sites/default/files/17/12/itdb-factsheet-2017.pdf, accessed 22/10/2018.
7. M. Higginson, P. Thompson, O. Marsden, S. Heath, F. Livens, L. Harwood and F. Lewis, Radiochimica Acta, 2015, 103(10), 687–694.
8. M. Srncik, K. Mayer, E. Hrnecek, M. Wallenius and Z. Varga, Radiochim. Acta, 2011, 99, 335–39.
9. R. Williams and A. Gaffney, Proc. Radiochim. Acta, 2011, 1, 31–35.
10. G. Eppich, R. Williams, A. Gaffney and K. Schorzman, J. Anal. At. Spectrom., 2013, 28, 666–674.
11. Z. Varga , A. Nicholl, M. Wallenius and K. Mayer, Anal. Chim. Acta, 2012, 718, 25–31.
12. M. Higginson, C. Gilligan, F. Taylor, D. Knight, P. Kaye, T. Shaw, P. Thompson, J. Radioanal. Nucl. Chem., 2018, 318, 157–164.
13. W. Davies, W. Gray, Talanta, 1964, 11, 1203–1211.
14. Nuclear Forensics International Technical Working Group (ITWG) Collaborative materials exchange exercise after action and lessons learned report 4. Coordinator: Schwantes J (29/10/2015) Pacific Northwest National Laboratory. PNNL-24410.
15. M. Kristo et al., J. Radioanal. Nucl. Chem., 2018, 315, 425.
16. www.mindat.org/loc-542.html, accessed 22/10/2018.
17. Z. Varga, M. Wallenius, K. Mayer and E. Hrnecek, Journal of Radioanalytical and Nuclear Chemistry, 2011, 290, 485–492.
18. M. Hubbard et al., PNNL Report 26680.
19. Nuclear Forensics International Technical Working Group (ITWG) Round Robin 3 Exercise after action and lessons learned report 3. Coordinator: Hanlen R (25/04/2011), Pacific Northwest National Laboratory. PNNL-20079.
20. T Kayzar and R Williams, J. Radioanal. Nucl. Chem., 2016, 307, 2061.

TIME-RESOLVED LASER-INDUCED LUMINESCENCE/CHEMILUMINESCENCE LASER SPECTROSCOPY AND DETECTION OF ACTINIDES/LANTHANIDES

I. N. Izosimov

Joint Institute for Nuclear Research, 141980 Dubna, Russia

izosimov@jinr.ru

ABSTRACT

This work is devoted to applications of the time-resolved laser-induced luminescence (TRLIF) spectroscopy and time-resolved laser-induced chemiluminescence (TRLIC) spectroscopy for detection of lanthanides and actinides. Results of the experiments on europium, samarium, uranium, and plutonium detection in different solutions are analysed.

1 INTRODUCTION

The use of laser radiation with tunable wavelength allows[1–4] selective excitation of actinide/lanthanide species with subsequent registration of luminescence or chemiluminescence for the detection of the elements. The practical application of laser spectroscopy to analysis of samples is confronted with one major difficulty, namely the element to be detected must be permanently located in the area of interaction with laser radiation. Therefore the use of solutions of the substances to be analysed is the most attractive from the practical standpoint. When the pulse (1ns) UV radiation produced by a nitrogen laser is used for lanthanide and actinide excitation in solutions the UV radiation is absorbed by many molecules and as a consequence the background radiation is increased. Using short laser pulses for excitation of molecules and ions in liquids and time resolution (TR) for registration of luminescence (TRLIF) and chemiluminescence (TRLIC) produced by actinide and lanthanide ions we can efficiently separate target signals from short-lived background luminescence.[1-6] Selective excitation of detectable molecules and multi-step excitation schemes of luminescence/chemiluminescence can additionally decrease the intensity of background radiation. The Limits of Detection (LOD) for spectrometers using the registration of chemiluminescence are in the range from 10^{-6} mol/l to 10^{-13} mol/l depending on the type of solutions and type of detectable molecule. Chemiluminescence is widely used as a detection method in many fields, such[7] as flow injection analysis, chromatography, biology, medicine, etc.

UV radiation is absorbed with chemiluminogen (luminol in our experiments) molecules, which makes difficult interpretation of the results of chemiluminescence registration. Therefore, a key problem of chemiluminescence application to detection of lanthanides and actinides in solutions is to increase in the selectivity of detection. Appropriate selectivity of lanthanide or actinide molecules excitation can be reached by initiation of transitions within 4*f*- or 5*f*-electron shells, which correspond to visible spectral range of absorbed laser radiation. Since the energy of one-quantum excitation in visible range may be insufficient for initiation of chemiluminescence it was proposed[1–4] to excite lanthanide or actinide ion by multi-quantum absorption of visible light. The scheme[1–4] *two step-one colour,* i.e. in irradiation of actinide-containing solution by one laser (two photons absorbed from one laser pulse) and the scheme *two step-two colour,* when a solution is irradiated by two lasers operating at different wavelengths (two photons absorbed from two synchronized laser pulses) were used for excitation of lanthanide/actinide ions in the range of 4*f*/5*f* electron transitions.

The details of multi-step excitation of luminescence/ chemiluminescence in solutions are considered. It is shown that a multi-step scheme of luminescence/chemiluminescence excitation increase both the sensitivity and selectivity of detection of substances.

2 TRLIF, SINGLE STEP EXCITATION SCHEME

The background radiation can be efficiently suppressed by using pulse laser radiation to excite solutions and measuring the luminescence with a delay (10^{-3} – 10^{-6} s) after the laser pulse. The solution to be analysed is exposed to a pulsed (10^{-8} – 10^{-9} s) laser beam. The luminescence spectrum is measured with a delay (10^{-3} – 10^{-6} s) with respect to the laser pulse (Figures 1 and.2). This method has been termed Time-Resolved Laser-Induced Fluorescence (TRLIF). The TRLIF method allows detection of lanthanide and actinide concentrations down[5] to 10^{-13} mol/l. The TRLIF technique features selectivity in four parameters: laser radiation wavelength, measured luminescence spectra (Figure 3), measured delay with respect to the pulse laser, and measured time (Figure 2). Also selectivity and sensitivity depend on the excitation scheme (single step or multi-step) and excitation wavelength[1–4]. For Eu, Sm, and U analysis we used a luminescence method with pulse (1ns) nitrogen laser excitation of the solution and time resolution for the signal registration (single step excitation TRLIF). One of the most convenient ways is the use of sodium polysilicate solution having a low self-background and providing a limit of uranyl detection in our experiments down to 0.005 ng/ml. However, this method is suitable only for analysis of inorganic samples. Biological samples containing a large amount of organic substances should be preliminary mineralized. Typical concentration of uranium in blood plasma is about 0.05 ng/ml – 0.5 ng/ml, in urine it is about 0.2 ng/ml – 5 ng/ml. Solution (2.2 ml) was placed into a quartz cuvette and the background luminescence was measured. Then, an aliquot of the solution to be analysed (in common cases, 0.05 ml – 0.2 ml) was added and the total intensity of background and the sample was determined. The decay time of uranium luminescence in polysilicate was approximately 500 µs. After mineralization of the sample and preparation of a solution for analysis the decay time of uranium luminescence was about 300 µs. The limit of detection decreases by a factor of 1.5-2 in passing of blood plasma added into solution from 0.05 ml to 0.15 ml – 0.25 ml. Thus, the permissible volume of blood plasma does not exceed 0.15 ml – 0.25 ml.

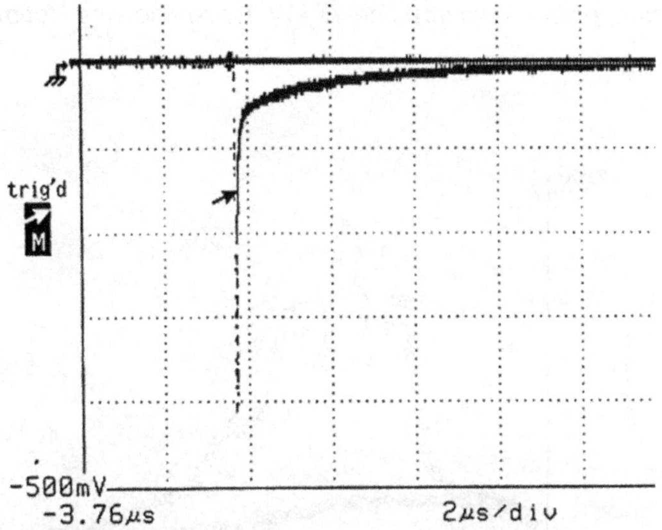

Figure 1 *Time dependence of the uranyl luminescence in a 4.5 M* H_2SO_4 *solution. Start is a nitrogen laser pulse. The short-lived background luminescence is clearly seen, as well as the relatively long-lived uranyl luminescence. The background luminescence can be significantly suppressed by measuring the luminescence with a delay of several µs after laser pulse.*

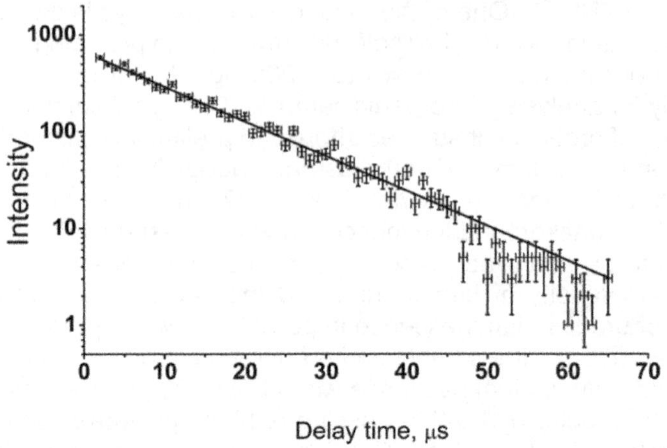

Figure 2 *Photoluminescence of $UO_2F_5^{3-}$ in {H_2O + CsF [42%]} solution, pH 9.0. Excitation by nitrogen pulse (10 ns) laser. Registration at λ=520 nm, δλ=9 nm. Gate time 1 μs. 200 laser pulses per channel were made. Laser beam diameter 5 mm, power in laser pulse 15 kW. Luminescence lifetime τ = 12.08 ± 0.25 μs.*

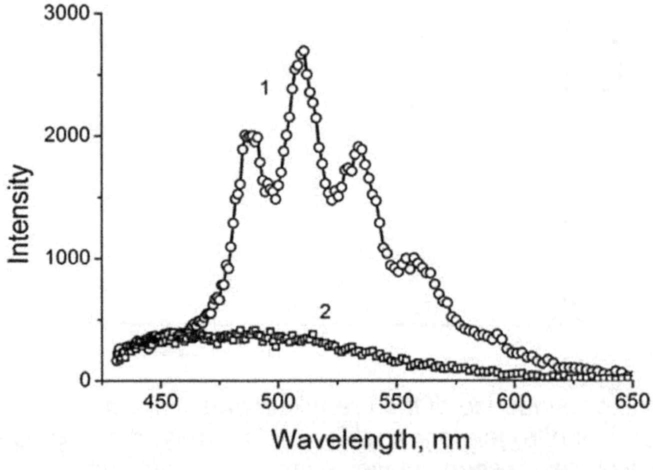

Figure 3 *Photoluminescence (1) of $UO_2F_5^{3-}$[0.005M] (○) in CsF(42%) + H_2O + luminol (10^{-4} M), and (2) luminol10^{-4}M (□) in CsF(42%) + H_2O, pH 10.07. Excitation by pulsed (10ns) nitrogen laser. Delay time 2μs. Gate time 100μs.*

Uranyl (or other elements and molecules) has a different circulation period in blood and in urine. By comparison of the concentrations of the uranyl (or other elements and molecules) in blood plasma and in urine one can estimate the time when the uranyl (or other elements and molecules) entered into the organism. Without mineralization the limit of uranyl detection in blood plasma was 0.1 ng/ml and after mineralization was up to 8 pg/ml – 10 pg/ml. The limit of uranyl detection in urine in our TRLIF experiments was up to 5 pg/ml. We applied TRLIF for samarium and europium detection in urine. We found that a high sensitivity of europium and samarium detection in aqueous solutions can be reached in the case of complex formation of these elements with fluorinated β-diketones and trioctylphosphine oxide (*TOPO*) in the presence of nonionic surfactants.

In this work, we used polytrifluoroacetone (*PTFA*), *TOPO*, and *Triton X–100*. The use of *PTFA* provides a low limit of detection of europium and samarium. The strongest luminescence radiation in the spectra of europium and samarium was observed at 614 nm and 643 nm, respectively. The wavelength of radiation maximum for both elements does not vary in passing from neat solution to a solution with the addition of urea solutions. The life-times of europium and samarium luminescence are 800 μs and 60μs, respectively, in both neat solution and solutions with addition of urine samples. By this it is meant that in this case there is no dynamic quenching of luminescence and the variation of luminescence intensity is apparently caused by absorption of exciting laser radiation in the solution in addition of urine sample. When 0.2 ml of urine is added into 2.2 ml solution the intensity of luminescence decreases by a factor of more than 2; hence, the volume of the sample required for analysis should be increased. The limit of detection was estimated from the 3σ background criterion, where σ is the standard deviation of the background measurements. In pure solution the limit of detection of europium was 0.005 ng/ml and samarium, 0.07 ng/ml. After addition of 0.2 ml of urine the limit of detection of europium was 0.015 ng/ml and samarium, 0.2 ng/ml.

Unfortunately, plutonium, neptunium, and also a number of valence forms of uranium give no direct luminescence in solutions. For determination of valence forms of plutonium, neptunium, and a number of valence forms of uranium other less sensitive methods of laser spectroscopy are used. Among them[1] are: laser induced photoacoustic spectroscopy (LIPAS) with LOD 10^{-7} M, absorption spectroscopy with LOD 10^{-5} M, and thermal lens laser spectroscopy (TLS) with LOD 10^{-6} M. We proposed to use high sensitive chemiluminescence method (TRLIC) for such actinides detection[1,8].

3 TRLIC, SINGLE STEP EXCITATION SCHEME

The behaviour of the UO_2^{2+} ion excited by radiation of pulse nitrogen laser in aqueous solutions with a high content of caesium fluoride and luminol was studied[1,9] at various pH's. Under irradiation by the nitrogen laser luminol gives luminescence in the same spectral range as the luminescence of uranyl ion (Figure 3). Naturally (Figures 3 and 4), there are two complexes $UO_2F_5^{\,3-}$ and

$UO_2F_4OH^{3-}$ in such solution[1,9]. $UO_2F_5{}^{3-}$ is the luminescent complex (Figure 3). $UO_2F_4OH^{3-}$ is a non-luminescent complex but after proper excitation it generates OH radicals, which can be detected from the enhancement of the luminol luminescence (Figure 4c).

Chemiluminescence of luminol under the action of OH radicals appearing in the solution was considered. In Figures 4a-4c are presented the kinetic data on luminol luminescence in uranyl-free and uranyl-containing solutions. It is evident that by increasing pH to 11.85 the ratio of the intensity of uranyl luminescence at the maximum of luminol luminescence in the solution containing uranyl increases in comparison with that of the solution containing no uranyl. An increase in pH results in the increase of the concentration of $UO_2F_4OH^{3-}$ complexes in the solution, an increase in the number of quanta absorbed by these complexes, and a decrease in the number of quanta absorbed by luminol molecules[1,9]. The decrease in the chemiluminescence efficiency with increasing pH of the solution from 8.19 to 9.86 (Figures 4a and b) suggests that the quantum efficiency of chemiluminescence under excitation of $UO_2F_4OH^{3-}$ is lower than that under optical excitation of luminol. The decrease in the luminol chemiluminescence owing to decrease in the light absorption of luminol with increasing pH is not compensated at this pH by chemiluminescence generated by excitation of $UO_2F_4OH^{3-}$. Further increase in pH (Figure 4c) results in increase in the total chemiluminescence efficiency.

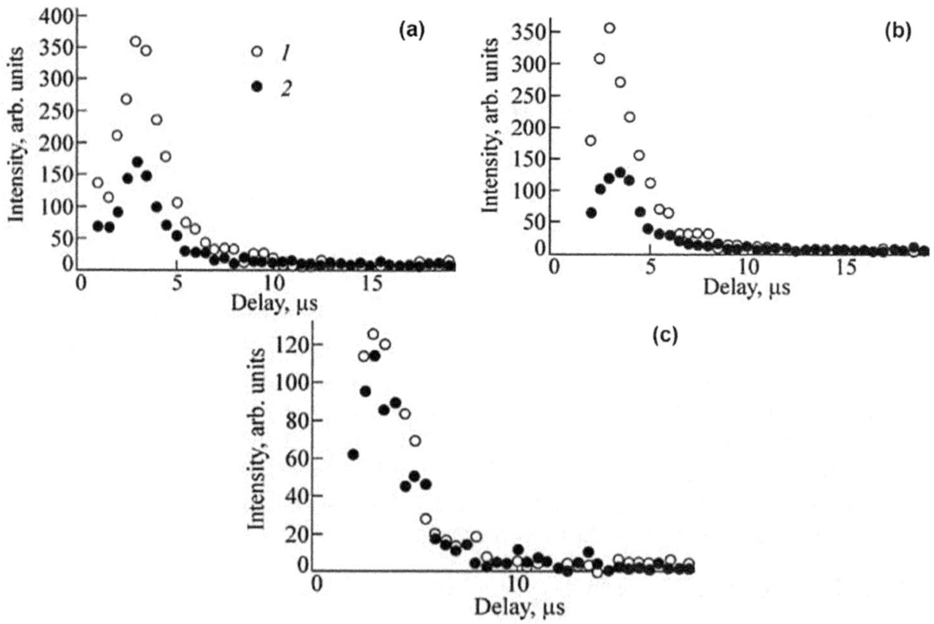

Figure 4 *Kinetics of luminol luminescence at the wavelength 459 nm in 42% solution of CsF in H_2O. pH: (a) 8.19. (b) 9.86. (c) 11.85. (1) - without uranium and (2) 0.009 M of $UO_2{}^{2+}$. Luminol concentration 10^{-4}. Gate time 1 μs.*

Data on luminol chemiluminescence in solutions containing complexes $AnO_2F_4OH^{3-}$ (An = U, Pu, Np) were[9-12] analysed. The luminescence was excited by nitrogen laser radiation. In the presence of uranyl, plutonyl, and neptunyl hydroxyfluoride complexes, luminol chemiluminescence sensitized by OH radicals was observed. Hydroxy radicals are generated by phototransfer of electrons from hydroxy ligand to the metal complex. The fact that the intensity of chemiluminescence initiated by photoexcited $AnO_2F_4OH^{3-}$ is comparable with that of chemiluminescence initiated after direct absorption of laser radiation with luminol molecules[9-11] indicates that the luminol chemiluminescence can be used to determine actinide traces in solutions.

The possibility of observation of chemiluminescence caused by the reaction of OH radicals with luminol molecules allows detecting the presence of actinyl ions UO_2^{2+}, PuO_2^{2+}, NpO_2^{2+}, and probably AmO_2^{2+} in the solution. The results we obtained[9-11] allowed a conclusion that absorption of UV radiation with UO_2^{2+}, NpO_2^{2+}, and PuO_2^{2+} ions forming mixed hydroxofluoride complexes under the experimental conditions leads to generation of OH radicals. Their formation results from an electronic transition with charge transfer from the hydroxide ion coordinated with AnO_2^{2+} to the actinide ion. Actually this is photochemical reduction of actinide ion[9-11]. The arising OH radicals initiate chemiluminescence, oxidizing luminol molecules. Thus, absorption of a laser pulse gives rise to a relatively long-lived chemiluminescence trace (Figures 4 and 5) allowing detection of actinide ions in solution using Time Resolved (TR) method[1,8]. This method allows detection of ions that cannot be detected by intrinsic luminescence.

We observed experimentally a time delay with respect to the laser pulse radiation and the maximum of the chemiluminescence pulse (building-up stage) for solutions containing luminol and actinide complexes (Figure 5). Such a delay is extremely significant since it allows application of the time resolution (TR) procedure for detection of chemiluminescence.

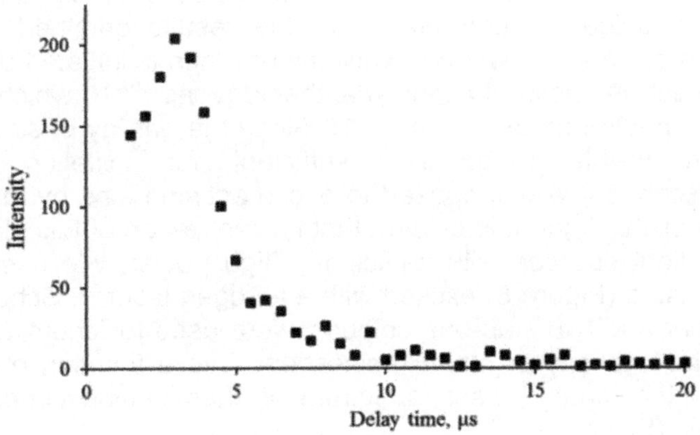

Figure 5 *Kinetic curve of luminol chemiluminescence induced by excitation of plutonyl ions in solution. Chemiluminescence building-up stage is about 3 μs.*

Chemiluminescence initiation by UV radiation of a nitrogen laser is unselective (Figure 6), and this does not allow identification of actinides and, the more so, of their valence forms[1,8]. A key problem of chemiluminescence application to the detection of lanthanides and actinides in solutions is an increase in the selectivity of detection.

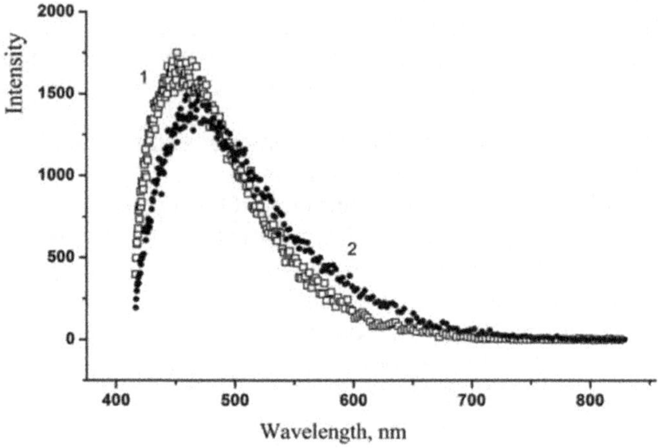

Figure 6 *Photoluminescence of $PuO_2F_5^{3-}$[0.001M] + luminol10^{-4}M (•) in CsF(42%) + H_2O (2), and (1) luminol10^{-4}M (□) in CsF(42%) + H_2O, pH 8.5. Excitation by pulsed (10ns) nitrogen laser. 200 laser pulse were made per channel at spectrum measurement, laser beam diameter 5mm, power in laser pulse 15kW. Delay time 2μs. Gate time 50μs.*

4 TRLIC, TWO STEP EXCITATION SCHEMES (TWO STEP-TWO COLOUR AND TWO STEP-ONE COLOUR)

Chemiluminescence initiation by radiation with $\lambda \leq 400$ nm is unselective (Figure 7). Furthermore, UV radiation is absorbed by luminol molecules, which additionally complicates interpretation of the results obtained. Appropriate selectivity can be reached when chemiluminescence is initiated by transitions within *5f/4f* electron shells of actinide/lanthanide ions[1-4,8,12], which correspond to the visible spectral range (Figure 7.1). Since the energy of single-quantum excitation in visible range is insufficient for initiation of luminol chemiluminescence it was proposed to excite actinide ions by multi-quantum absorption of visible light. It is evident that for realization of this idea there is a need to use light sources with sufficiently high power. We used two pulse tunable dye lasers (Figure 8) excited with a nitrogen laser[1-4]. Schemes for two step-two colour and two step-one colour[1-4] were used for chemiluminescence excitation. The intensity of chemiluminescence as a function of wavelength generated by the tunable laser (spectrum of chemiluminescence excitation) was measured (Figures 9–12).

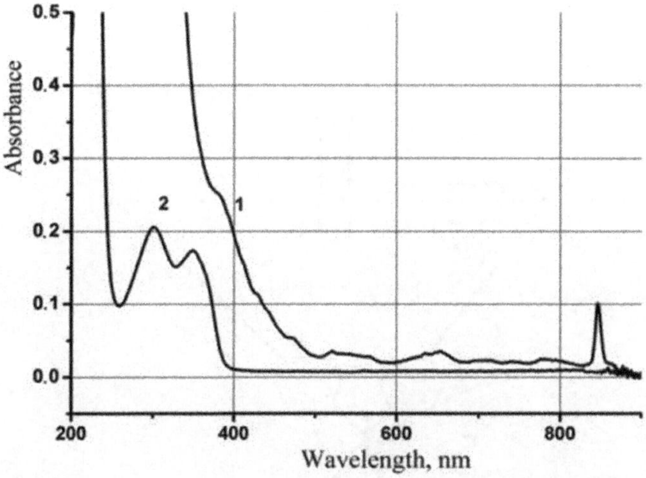

Figure 7 *Absorption spectra : 1 - PuO_2^{2+} (0.003 M) in solution 42% CsF+H_2O; 2 - luminol [0.001 M] in solution 42% CsF+H_2O, pH 10.5.*

The experiments were performed on an installation (Figure 8) consisting of pulse nitrogen laser OBB 1010 with a pulse length of 1ns and generation power of approximately 1.4 MW and two dye lasers OBB 1012 and OBB 1011. When using two dye lasers the radiation generated by the nitrogen laser was simultaneously divided to both dye lasers through a beam splitter. This scheme allows synchronization of laser pulses in a cuvette within an accuracy of 10 ps at generation pulse length of 800 ps for laser OBB 1012 and 1 ns for laser OBB 1011. The laser beam splitter was oriented at an angle of 45° to the direction of laser beam generated by the nitrogen laser and divided this beam into two beams with equal intensities. Laser beams generated by two dye lasers were aligned in the opposite directions and directed to a cuvette 1 cm in thickness. OBB's 1012 Dye Laser incorporates a grazing incident design laser cavity for high resolution followed by a secondary amplifier cell to boost the power. The result is a narrow 0.04 nanometer output from 360 nm to 900 nm, a pulse width of 1 nanosecond, and an energy of 220 microjoules per pulse at 500 nm. With the addition of OBB's OL-403 Frequency Doubler, tunable wavelengths from 235nm to 345 nm can be attained.

Chemiluminescence of luminol was collected with a lens whose optical axis was oriented at an angle of 39° to the direction of the laser beams and was transferred to the entrance slit of a double prismatic monochromator DMR-4 with flexible optical fibre.

Chemiluminescence was recorded in the quantum counting mode[1-3] with the use of a gating technique at a wavelength of 460nm corresponding to the maximum of luminol chemiluminescence. The length of gating impulse (strobe) was 10 μs, delay time, 2 μs.

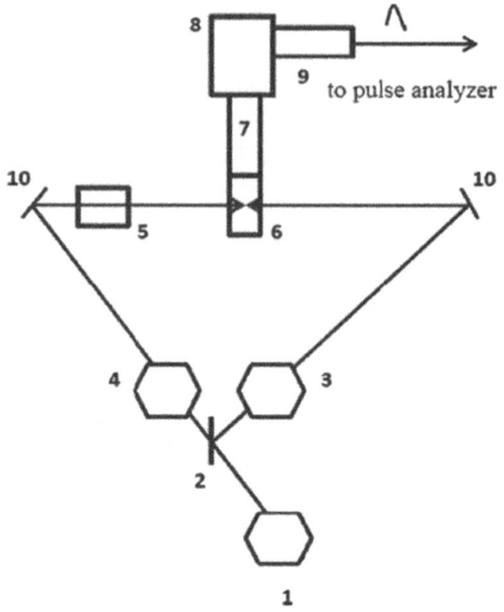

Figure 8 *Scheme of the experimental set up: (1) nitrogen laser OBB-1010, (2) beam splitter, (3) dye laser OBB-1011, (4) dye laser OBB-1012, (5) optical delay line OPD-1, (6) cuvette with solution, (7) optical fibre, (8) monochromator DMR-4, (9) photomultiplier, (10) mirror.*

A radiation wavelength (λ_1) of laser OBB 1012 was tuned in the limits of absorption band of detectable actinide valence or molecular form. A radiation wavelength of laser OBB 1011 was fixed in the region λ_2=490nm–500nm (two colour-two step scheme of chemiluminescence excitation). In one step – one colour (two laser photons absorbed from one laser pulse) only the tunable OBB 1012 (λ_1) laser was used. The intensity of chemiluminescence as a function of wavelength generated by the tunable laser (spectrum of chemiluminescence excitation) was measured. The presence of absorption bands of detectable actinide in the range of tuning of the first (λ_1) laser wavelength results in appearance of a peak of luminol chemiluminescence. Peaks in the spectrum of chemiluminescence excitation are connected with the definite valence or molecular forms of detectable actinides. The intensity of chemiluminescence was measured[2-4] only during the strobe pulse duration with the proper delay time after laser pulse (TRLIC). Thus, background radiation can be efficiently suppressed and chemiluminescence signal will be much sharper.

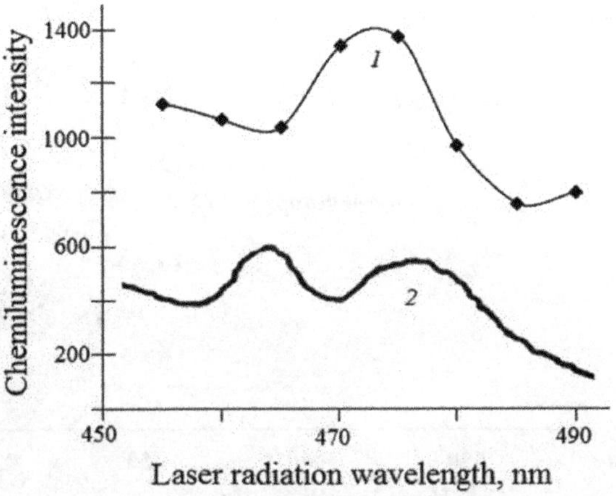

Figure 9 *(1) Spectrum of chemiluminescence excitation by the scheme two step-one colour in luminol + Sm(III) solution. (2) Absorption spectrum of Sm^{3+}.*

The experiments were carried out at concentrations of *f* element of the order of 10^{-3}M. It is shown that a multi-step scheme of chemiluminescence excitation increases the selectivity of the *f* element detection. Because the LOD for the chemiluminescence method of detection may[7] reach 10^{-13} M, the multi-step excitation scheme for the chemiluminescence initiation is promising for further development of a method that may become competitive in sensitivity and selectivity with Inductively Coupled Plasma Mass Spectrometry (ICP-MS) and other trace detection methods[1].

We recorded the spectra of chemiluminescence excitation as a result of excitation of Sm^{3+} ions with dye laser by using two step -one colour scheme[2-4]. There is no complete similarity of the spectrum of chemiluminescence excitation to absorption spectrum (Figure 9). This experimental fact is connected with the difference in the selection rule for single-quantum and multi-quantum absorption[2-4].

Figure 10 shows a portion of the U(IV) absorption spectrum and the chemiluminescence intensity varying with changes in the generation wavelength of the OBB 1012 tunable laser. As can be seen, the chemiluminescence excitation spectrum on varying the generation wavelength of the first laser is similar to the uranium absorption spectrum in the tuning range. The presence of the U(IV) absorption band in the region of tuning of the emission wavelength of the first laser gives rise to a peak of the luminol chemiluminescence intensity. This fact undoubtedly reflects the selective mechanism of the chemiluminescence excitation[2-4].

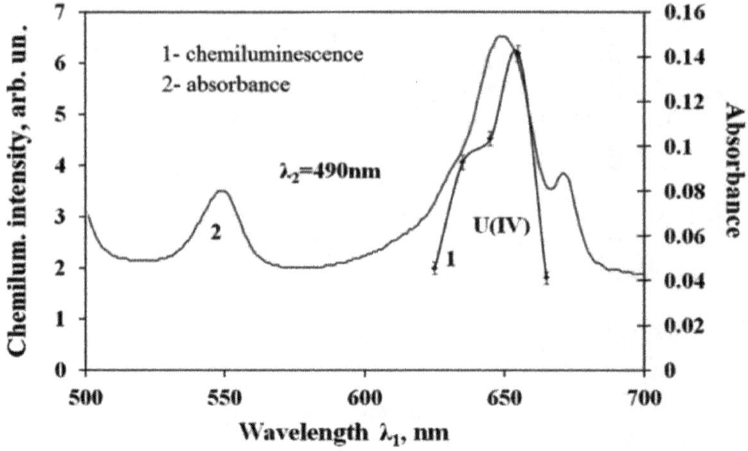

Figure 10 *(1) Spectrum of chemiluminescence excitation by the scheme two step-two colour in luminol + U(IV) solution. (2) Absorption spectrum of U(IV) in aqueous HCl solution.*

Initiation of chemiluminescence as a result of excitation of Pu(IV) with one (Figure 11) and two (Figure 12) dye lasers was demonstrated for a solution containing caesium fluoride, luminol, and Pu(IV). A solution composition was chosen[2-4] in such a way as to provide favourable conditions for observation of luminol chemiluminescence and to avoid formation of colloidal species of hydrolysed Pu(IV). As seen the spectrum of chemiluminescence excitation is in close agreement with absorption spectrum of Pu(IV) which indicates high selectivity of chemiluminescence excitation. A comparison of the curves presented in Figures 11 and 12 shows that in the mechanism two step – one colour (Figure 11) the slop of the spectral curves is sharper than that in the mechanism two step – two colour (Figure 12). The spectrum of chemiluminescence excitation correlated with absorption spectrum of Pu(IV). In both schemes we realized selective excitation of chemiluminescence, and this selectivity is caused by the features of absorption spectra of Pu(IV) solutions.

A measurement of the spectrum of chemiluminescence excitation requires correct consideration of the contributions from the following processes.

1. Initiation of luminol chemiluminescence as a result of two-quantum excitation of An(IV) by the scheme two step one colour, i.e. as a result of absorption by An(IV) of two quanta radiated by one laser.

2. Initiation of luminol chemiluminescence as a result of two-quantum excitation of An(IV) by the scheme two step - two colour, i.e. as a result of absorption by An(IV) of two quanta radiated by two lasers.

3. Initiation of luminol chemiluminescence as a result of two-quantum excitation of luminol molecules.

Of course, the process (3) is a background process and it's spectrum of chemiluminescence initiation does not correlate with the absorption spectrum of An(IV).

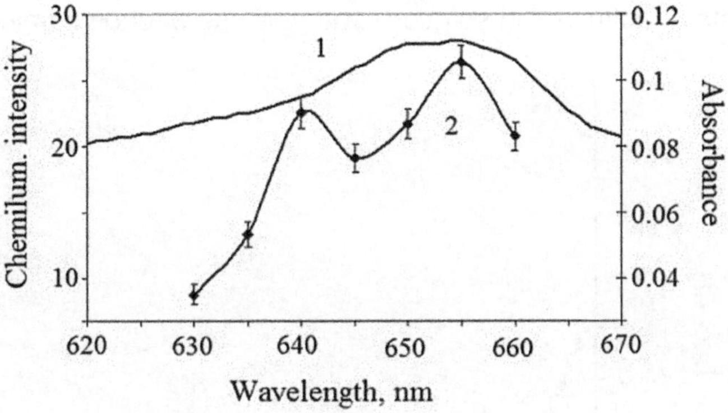

Figure 11 *(1) The absorption spectrum of Pu(IV) in solution. (2) Chemiluminescence excitation spectrum for the luminol + Pu(IV)+CsF solution using the two step – one colour scheme.*

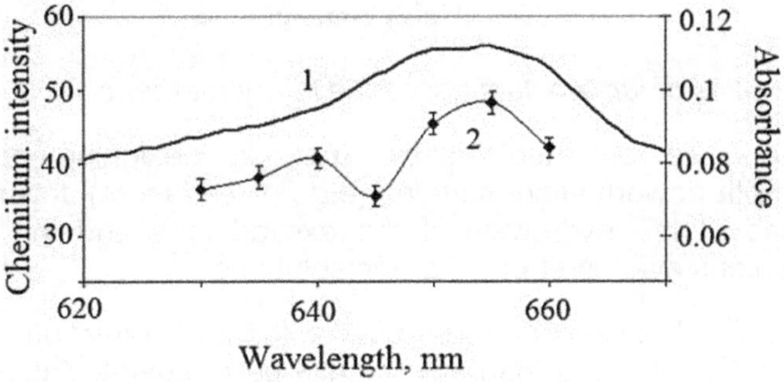

Figure 12 *Comparison of (1) the absorption spectrum of Pu(IV) and (2) the intensity of chemiluminescence at various wavelengths of radiation generated by the OBB-1012 dye laser (two step – two colour scheme).*

For detection of actinides with the use of chemiluminescence recording it is necessary to exclude the possibility of luminol luminescence having a nature different from chemiluminescence. In this connection we studied the kinetics of this luminescence in alcoholic solutions with different water content. It was found that in single-quantum excitation with decreasing water content the intensity of chemiluminescence decreases and luminescence, having no (Figure 13) the building-up stage typical for chemiluminescence (Figure 5), becomes more pronounced. It should be noted that this luminescence (Figure 13), different from chemiluminescence arises, (Figure 5) in single-quantum UV excitation of luminol molecules and can be significantly depressed in two-quantum excitation induced by radiation with longer wavelength, since in the visible region a scheme of two-quantum excitation involving no absorption band of luminol can be chosen (Figure 7). We showed that this background luminescence of luminol can be significantly depressed in two-quantum excitation of solutions by laser radiation with wavelengths longer than 450 nm. This fact is very important in choice of background conditions for recording of trace amounts of actinides in solutions with the use of chemiluminescence.

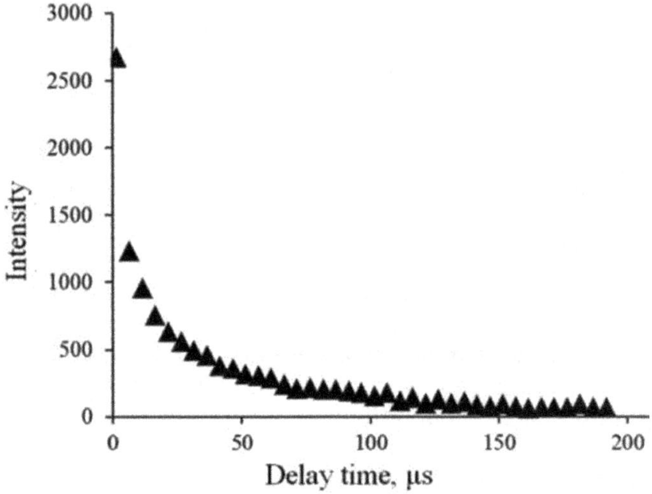

Figure 13 *Kinetics of luminol luminescence in dry methanol.*

The results we obtained[13] show that, in designing photochemical experiments with powerful light sources (e.g., pulse lasers), it is necessary to take into account the hydrolysis of the excited ions and the concomitant chemical reactions even in strongly acidic solutions.

Also, because of the second order effects, the laser radiation should not be too powerful[2-4] (usually less than 10^8 W cm^{-2} per a pulse). Thus, when using the multi-step schemes of chemiluminescence excitation there is a need to choose the power of laser radiation to provide the required sensitivity and selectivity.

5 DISCUSSIONS AND CONCLUSIONS

At present the most efficient methods of detection of actinides and lanthanides in solutions are methods based on registration of actinides with time resolution, time-resolved laser induced fluorescence (TRLIF) spectroscopy, having[1,5] a limit of detection (LOD) up to 10^{-13} M. TRLIF may be applied for biological sample analysis. Comparing concentrations of the uranyl (or other elements and molecules) in blood plasma and in urine one can estimate the time when the uranyl (or other elements and molecules) got into an organism. Samples containing a large amount of organic substances should be preliminary mineralized.

Plutonium, neptunium, and some uranium compounds do not produce direct luminescence in solutions, but when excited by laser radiation, they can induce[1-4] chemiluminescence of chemiluminogen (luminol in our experiments). It is because of its high sensitivity (LOD from 10^{-6} M to 10^{-13} M) that chemiluminescence is widely used in many[7] fields. A key problem of chemiluminescence application to detection of lanthanides and actinides in solutions is an increase in the selectivity of detection. Appropriate selectivity can be reached when chemiluminescence is initiated by transitions within *4f/5f* electron shell of lanthanide/actinide ions, which correspond to the visible spectral range. In some cases chemiluminescence can be simpler to detect than intrinsic luminescence. For example[14], in U(IV) solutions it is possible to excite luminescent transitions, but their lifetimes are shorter than 1 ns, and recording of such transitions is difficult. Since the energy of one quantum excitation in visible range is insufficient for initiation of luminol chemiluminescence it was proposed[1-4,8] to excite lanthanide/actinide ions by multi-quantum absorption of laser radiation with wavelengths longer than 450 nm.

The selective excitation of actinide gives rise a chemical reaction between molecule or complex containing excited actinide and chemiluminescence agent added into solution. As a result of the reaction the light is emitted (chemiluminescence) and registered. Using laser radiation with tunable wavelength we can selectively excite various valence forms and molecules of actinides with subsequent registration of chemiluminescence. With knowledge of wavelength at which chemiluminescence appears and the intensity of chemiluminescence we can determine the concentration of a certain valence form of given actinide and the structure of complex containing this actinide[1-4,8]. The optimum choice of laser radiation wavelength on the basis of absorption spectra, scheme of luminescence/chemiluminescence excitation (one-step or multi-step), and chemiluminescence agent is extremely significant for selective and efficient luminescence induction in excitation of actinides. It should be noted that the presence of a time delay between the pulse of laser radiation and the chemiluminescence pulse is extremely significant feature[1,8]. This fact allows using the time resolution (TR) procedure for detection of chemiluminescence (TRLIC). TRLIC method allows one to obtain additional information about actinides/lanthanides compare with ICP-MS. The main limitation of ICP-MS methods is their inability to distinguish between molecular and isotopic isobars, which complicates the analysis. Apply the laser radiation with tunable wavelength, one can selectively excite various valence forms and molecules of actinides/lanthanides with subsequent registration of chemiluminescence. When knowing the wavelength at which chemiluminescence appears and the intensity of this chemiluminescence, one can determine the concentration of a certain valence form of a given actinide/lanthanide and the structure of the complex containing this actinide/lanthanide.

Using chemiluminescence of solutions, we found an approach[2-4] to registration of plutonium, uranium, samarium and other elements in solutions with a high sensitivity in excitation of plutonium, uranium and samarium with a pulse laser with tunable wavelength. A multi-step scheme of chemiluminescence excitation makes this procedure not only highly sensitive but also highly selective method of detection of substances.

CONFLICTS OF INTEREST

There are no conflicts to declare.

REFERENCES

1. I. N. Izosimov, *Phys. Part. Nucl.*, 2007, **38**, 177.
2. I. N. Izosimov, N. G. Firsin, N. G. Gorshkov and S. N. Nekhoroshkov, *Hyperfine Interactions*, 2014, **227**, 271.
3. I. N. Izosimov, *J. Radioanal. Nucl. Chem.*, 2015, **304**, 207.
4. I. N. Izosimov, *Procedia Chemistry*, 2016, **21**, 473 (2016).
5. C. Moulin, P. Decambox and P. Mauchien, *J. Radioanal. Nucl. Chem.*, 1997, **226**, 135.
6. C. P. Baird and T. J. Kemp, *Prog. React. Kinet.*, 1997, **22**, 87.
7. C. Dodeigne, L. Thunus and R. Lejeune, *Talanta*, 2000, **51**, 415.
8. I. N. Izosimov, N. G. Gorshkov, L. G. Mashirov, *et al.*, in: *Proc. Int. Conf. "Actinides-2005"*, Manchester, UK, p.779.
9. N. G. Gorshkov, I. N. Izosimov, A. A. Kazimov, *et al.*, *Radiochemistry*, 2001, **43**, 354.
10. N. G. Gorshkov, I. N. Izosimov, A. A. Kazimov, *et al.*, *Radiochemistry*, 2001, **43**, 360.
11. N. G. Gorshkov, I. N. Izosimov, A. A. Kazimov, *et al.*, *Radiochemistry*, 2003, **45**, 28.
12. N. G. Gorshkov, I. N. Izosimov, A. A. Kazimov, *et al.*, *Optics and Spectroscopy*, 2002, **92**, 182.
13. N. G. Gorshkov, I. N. Izosimov, V. A. Mikhalev, *et al.*, *Radiochemistry,* 2011, **53**, 168.
14. N. G. Gorshkov, I. N. Izosimov, V. A. Mikhalev, *et al.*, *Radiochemistry*, 2012, **54**, 525.

CHARACTERISATION OF FIR1 TRIGA MARK II RESEARCH REACTOR – A COMBINATION OF MODELLING AND EXPERIMENTAL

Anumaija Leskinen, Antti Räty and Markus Airila

VTT Technical research centre of Finland Ltd

Anumaija.Leskinen@vtt.fi

ABSTRACT

Characterisation of FiR1 TRIGA Mark II research reactor in Finland relies on a combination of modelling and experimental results. Modelling is used in order to form an estimation of material specific activation levels. This information is used, for example, in planning of the dismantling process and waste management. Since some uncertainties are present in modelling results, experimental verification of activity levels are needed. Non-destructive analysis (NDA) using gamma spectrometry can be carried out for gamma emitting radionuclides, whereas analysis of alpha and beta emitting radionuclides, termed also difficult to measure (DTM) radionuclides, requires chemical separations and purification of the radionuclide of interest. Due to the lengthy and costly radiochemical analysis of DTM, a scaling factor is often determined to link the gamma emitting radionuclides with DTM. For FiR1, scaling factors are determined prior to the active dismantling phase and thus activity and nuclide vector calculations of waste packages can be carried out during, or even after, the active dismantling phase.

1 INTRODUCTION

Decommissioning of power and research reactors is a rising topic worldwide. While some countries already have well-established decommissioning regimes, some others are in an initial decommissioning phase or are moving towards decommissioning in the near future. Finland is in the initial decommissioning phase of its first reactor, namely FiR1, which is a TRIGA Mark II type research reactor in the greater Helsinki area. After its commissioning in 1962, it served the training of nuclear energy and technology professionals for over 50 years until its final shutdown in 2015. The early operations included intensive neutron beam research, activation analyses and isotope production. In the 1990s, a Boron Neutron Capture Therapy (BNCT) treatment facility was constructed and connected to the reactor and over 200 cancer patients were treated. Currently, a decommissioning licence has been applied for from the government and the active dismantling phase is planned for early 2020s.

Dismantling of FiR1 will produce radioactive waste packages containing activated reactor materials. In comparison to nuclear power plant decommissioning, the amount of activated materials from FiR1 will be much smaller in volume and radioactivity, whereas the variety of materials will be remarkable, especially due to the BNCT station. The largest volume of waste will be the activated parts of the concrete biological shield. Other materials will be graphite, steel, lead, Boral (shielding material for thermal neutron made of an aluminium boron alloy), aluminium, Fluental™ (moderator material made of an aluminium fluoride and lithium fluoride mixture) moderator material, and lithium enriched neutron shielding plastics. Therefore, characterisation of a variety of activated materials is needed for safe and cost-efficient dismantling planning and execution.

2 CHARACTERISATION

The characterisation of FiR1 relies on modelling and experimental efforts, which feed input to each other and therefore reinforce the overall characterisation process. Initial planning of the dismantling process, development of sampling plan of the activated reactor components and designing of material specific waste packages can be carried out after modelling of reactor component specific activation levels. After having an estimation of the activation levels and radionuclide vectors, experimental analyses can be carried out at relevant locations. Also missing data on chemical compositions of the un-activated reactor materials, such as trace level cobalt concentration in concrete, can be determined experimentally. The experimentally determined data is used to reinforce the modelling results.

Characterisation is also used in determination of scaling factors. Scaling factors are material and reactor specific and they essentially contain the assumption that a waste stream is stable enough to be described by constant ratios between radionuclide concentrations (forming so-called radionuclide vectors for each waste stream). More specifically, scaling factors are formed by linking gamma-emitting radionuclides with alpha/beta-emitting radionuclides, which are often referred to as difficult to measure (DTM) radionuclides.[1] Analysis of DTM radionuclides often requires lengthy radiochemical separations and purifications, whereas gamma-emitting radionuclides can be determined using non-destructive gamma spectrometric analysis. Therefore, it is beneficial to form scaling factors in pre-dismantling phase in order to analyse only gamma-emitting radionuclides and calculate DTM radionuclides during the dismantling phase.

3 RESULTS AND DISCUSSION

The initial modelling of the FiR1 was carried out in three steps. The first step consisted of collecting input data on reactor geometry, operating history and the chemical compositions of structural materials. The second step consisted of the calculation of neutron fluxes using a 3D reactor model built with Monte Carlo code MCNP. The third step included the combining of data on fluxes with operating history using point kinetic code ORIGEN-S. The modelling results including component level radionuclide inventory and activation results are presented in a VTT research report.[2] In these results, conservative assumptions were used for missing data on detailed chemical compositions of the materials, and partly missing operational data. In-house analytical capabilities were utilised as much as possible in the analysis of chemical compositions and especially HR-ICP-MS was used in the analysis of minor impurities, which may be efficiently activated under neutron irradiation and thus have a significant effect on the overall activity level of the material (for example lithium in concrete is efficiently activated via $^{9}Li(n,\alpha)^{3}H$). The results were also compared with available data on other decommissioned research reactors. Preliminary waste characterisation measurements have been reported earlier.[3]

Up to date, radiochemical separation methods have been tested for ^{3}H, ^{14}C, ^{36}Cl, ^{55}Fe and ^{63}Ni analysis in graphite, lead, aluminium, steel, and spent ion exchange resins. Low and very low-level samples have been analysed in order to gain confidence in the separation methods, but higher activity samples are needed for the calculation of reliable scaling factors.

CONCLUSIONS

Characterisation of FiR1 reactor relies on modelling and experimental results. The modelling results have given valuable input into the decommissioning planning and currently the material wise nuclide vectors and activation levels are verified experimentally. In order to complete the dismantling project efficiently, scaling factors will be determined in the near future.

CONFLICTS OF INTEREST

There are no conflicts to declare.

REFERNCES

1. *IAEA-TECDOC-1537*, 2007.
2. P. Kotiluoto and A. Räty, *VTT Research Report VTT-R-03599-16*, 2016.
3. A. Räty, T. Kekki, M. Tanhua-Tyrkkö, T. Lavonen and E. Myllykylä, *Nuclear Technology*, 2018, **203**, 205.

SIMPLIFIED METHOD FOR DETERMINING RESIDUAL SPECIFIC ACTIVITY IN ACTIVATED CONCRETE OF A PET-CYCLOTRON ROOM USING A SURVEY METER

Hiroshi Matsumura,*[a] Go Yoshida,[a] Akihiro Toyoda,[a] Kazuyoshi Masumoto,[a] Koichi Nishikawa,[a] Takayuki Nakabayashi,[b] Yoshiharu Miyazaki,[c] Taichi Miura,[a] Hajime Nakamura[a] and Kotaro Bessho[a]

[a] High Energy Accelerator Organization (KEK), 1-1 Oho, Tsukuba, Ibaraki, 305-0801 Japan. [b] Japan Environment Research Co., Ltd., 6-24-1 Nishi-Shinjuku, Shinjuku-ku, Tokyo 160-0023, Japan. [c] The Medical and Pharmacological Research Center Foundation, Wo-32, Inoyama-machi, Hakui, Ishikawa 925-0613, Japan.

E-mail: hiroshi.matsumura@kek.jp

ABSTRACT

An in situ method to determine the specific activity in concrete structures using a survey meter was established in this study for decommissioning accelerator rooms. This method is much simpler than the conventional sampling method that requires core boring. A survey meter (T-SP2, TAC Inc., Japan) was customised and equipped with a small detector probe, whose size allowed the design of a light-weight Pb shield. For dose rate measurement, a Pb shield thickness of 4.5 cm was enough to shield ambient γ-rays, and its weight (16 kg) was low enough for the device to be carried by a human. A calibration curve was obtained to quickly and easily convert the net contact dose rate on concrete into the specific activity in a PET cyclotron room. The specific activity detection limit for this device was lower than the clearance limit of ^{152}Eu and ^{60}Co in activated materials. Therefore, this method can be used for separating non-activated concrete from activated concrete during decommissioning. A general correction factor for the slope of the calibration curve was also obtained for use at other accelerator facilities.

1 INTRODUCTION

Many accelerator facilities in Japan will soon reach the end of their useful life and be considered for decommission. When decommissioning an accelerator facility, the activated parts of concrete should be separated from the non-activated parts in order to minimise the amount of activated waste. For efficient decommissioning, a simple method is required to determine the residual specific activity within the very large volumes of concrete walls, floor, and ceiling of the accelerator room.

Conventionally, the specific activity in concrete is determined by a sampling method using core boring.[1-3] This method is able to provide accurate specific activity as well as information on its depth profile. However, the sampling method has four significant disadvantages for the decommissioning work: (1) the local point determination for the large concrete faces, (2) the frequency of core boring is limited to e.g. 1 or 2 cores on each concrete face, (3) destructive measurement of the activated concrete, and (4) the requirement of off-site γ–ray measurement. In summary, the sampling method requires an immense amount of time and effort to determine the specific activity.

In general, tritium activity is highest among the long lived radionuclides in the concrete of the accelerator facilities.[2] However, γ–ray emitting radionuclides such as ^{152}Eu and ^{60}Co are much more important than tritium for decommissioning of the concrete because the clearance limits of ^{152}Eu and ^{60}Co are 1/1000 of tritium in Japan. Therefore, destructive measurement by sampling is not required for the tritium determination.

By bringing a detector into the accelerator room, in situ measurement bypasses the disadvantages of the sampling method to quickly determine and map the specific activity. In a previous study,[4] we examined an in situ determination method by contacting a portable Ge detector with the concrete surfaces of a positron emission tomography (PET) cyclotron room. The obtained results were compared with those from the sampling method, confirming successful measurement of the specific activity by the in situ method using the portable Ge detector.

The same study[4] also pointed out that a survey meter can replace the portable Ge detector as an easier and speedier alternative. As the next step, in this follow-up study we establish an in situ determination method using a survey meter.

2 EXPERIMENTAL

2.1 PET cyclotron room as experimental site

The PET cyclotron room of the Medical and Pharmacological Research Center Foundation (Hakui, Ishikawa, Japan) was used as the experimental site, the same as in the previous study.[4] Detailed information about the facility (cyclotron performance, operation history, room dimension, etc.) was summarised by Matsumura et al.[4] An NKK-Oxford superconducting cyclotron[5] had been used there to produce radiopharmaceuticals for PET. After operation from 1998 to 2015, the entire cyclotron body and its accessories were removed from the room. The present experiment was performed from March 2018 to May 2018, approximately 3 years from the end of operation.

2.2 Specific activity determination in concrete walls from previous work

In the previous study,[4] four depth profiles of the specific activity in the different four walls of the accelerator room were obtained by the sampling method. The distribution of the specific activity was also measured by the in situ method using the portable Ge detector at 20 locations.

By the sampling method, five long half-life nuclides, ^{152}Eu, ^{60}Co, ^{134}Cs, ^{154}Eu, and ^{54}Mn, were detected in the concrete walls approximately 2 years after the end of operation. The dominant radionuclides were found to be ^{152}Eu and ^{60}Co for the decommissioning of the concrete walls.

Since all of the depth profile of the specific activity was uniform from the surface to a depth of approximately 20 cm, the specific activity could be determined by the in situ method using a portable Ge detector. In the walls of the accelerator room, there were 10 cm mesh grids of reinforcing rods 1.91 cm in diameter, placed at a depth of 5 cm from the concrete surface. In the in situ measurement, the ^{60}Co produced in the reinforcing rods affected to the determination of ^{60}Co in the concrete walls. The counting contribution of ^{60}Co γ–rays from the reinforcing rods was estimated to be 54.7% of all. Then, the specific activities were corrected for the contribution from the reinforcing rods.

Furthermore, it was found that the specific activity ratios of each radionuclide against ^{152}Eu were independent of the location on the walls. These facts indicate that the contact dose rate at the concrete wall is proportional to the specific activity within the facility.

2.3 In situ determination of specific activity in concrete walls and floor using a portable Ge detector

In this study, we newly measured the specific activities of ^{152}Eu and ^{60}Co at 19 locations on the walls and at 14 locations on the floor by using a portable Ge detector with a 6.5 cm thick Pb shield. The determination procedure was the same as in the previous work.[4] In the floor, 20 cm mesh grids of reinforcing rods with 1.27 cm diameter had been placed at a depth of 5 cm from the concrete surface. From the differences of mesh density and diameter of the reinforcing rods, the density of these reinforcing rods in the floor was simply estimated to be 1/4.5 of that in the wall. Therefore, in the case of the floor, the counting contribution of ^{60}Co from the reinforcing rods was estimated to be 21.1%, which was used to correct the measured specific activities of concrete.

2.4 Customised survey meter equipped with a small detector probe

For in situ measurement of dose rate using a survey meter, the detector probe must be shielded from ambient γ-rays by a Pb shield. For portable use, the Pb shield should be as light as possible, which requires minimisation of the detector probe within the shield.

The survey meter is usually equipped with a long cylindrical shaped detector probe containing a scintillator combined with a photomultiplier tube (PMT). As a typical survey meter of the NaI scintillator + PMT type, TCS-171 (Hitachi Aloka Medical, Ltd., Japan, shown in the right side of Figure 1) has a probe dimension of 3.7 cm in diameter and 21.5 cm in length. The length of the PMT could not be shortened due to performance degradation. Therefore, TSC-171 required a large and heavy Pb shield.

To minimise the detector probe, a CsI scintillation survey meter (T-SP2, TAC Inc., Japan) adopting multi-pixel photon counter (MPPC) instead of PMT was customised for this study. Figure 1 shows photographs of the customised T-SP2 on the left side. This detector probe is much smaller than that of TCS-171.

The customised T-SP2 consists of three parts: the detector probe, the signal circuit, and the analyser. In the original T-SP2, the detector probe and signal circuit were packaged in a box. Since the 1-inch-cubic CsI scintillator combined with MPPC was separated from the signal circuit, the size of the detector probe was only 4 cm × 5 cm × 3 cm.

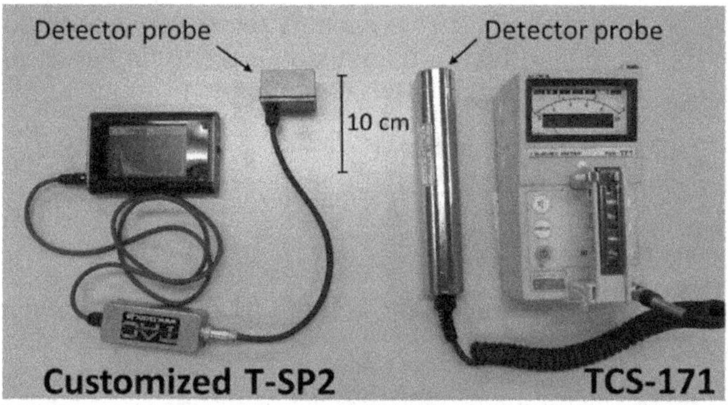

Figure 1 *Photographs of the T-SP2 CsI-scintillation survey meter customised for this study (TAC Inc., Japan) and the TCS-171 NaI-scintillation survey meter (Hitachi Aloka Medical, Ltd., Japan) as an example of the typical survey meter.*

2.5 Test Pb shields for the customised T-SP2

In order to make the shield as light as possible, we examined the shielding effect from ambient dose rate for different Pb thicknesses. For this purpose, 1, 2.5, 4.5, and 6.5 cm thick Pb shields with a space just enough to fit the detector probe of the customised T-SP2 were made. Figure 2 shows the design of the 4.5 cm thick Pb shield. The open mouth for installing the detector probe has an area of 4.5 cm × 5.9 cm, which is the same for all Pb shields. Figure 3 is a photograph of the Pb shield installed onto the customised T-SP2. There is a small hole for a cable path on the back of the Pb shield. The detector probe is placed in the open space at the bottom, and the CsI scintillator crystal placed in an invisible position through the hole. The Pb shield is divided into two bodies, and each body has a carry handle.

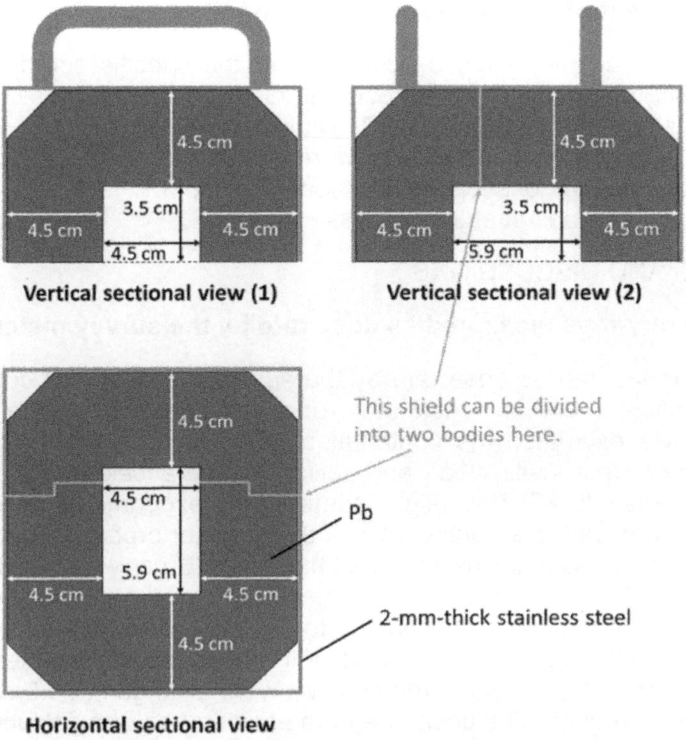

Figure 2 *Example vertical and horizontal sectional views of the 4.5 cm thick Pb shield for the customised T-SP2.*

Figure 3 *Photographs of the 4.5 cm thick Pb shield installed onto the customised T-SP-2.*

2.6 Measurement of contact dose rates using the customised T-SP2

At the same locations on the concrete where the specific activity had been determined in this study using the portable Ge detector, the dose rates were measured using the customised T-SP2 without or with 1, 2.5, 4.5, and 6.5 cm thick Pb shields. The measurements were performed at 19 locations on the concrete walls and 14 locations on the concrete floor of the accelerator room. The time constant was set at 30 seconds.

3 RESULTS AND DISCUSSIONS

3.1 Source of γ-rays measured as dose rate by the survey meter

The γ-rays measured as dose rate by the survey meter should come mainly from the concrete of the accelerator room it contacts. Using a simplified calculation, we estimated the contribution ratios of dose rates from the six concrete faces (four walls, floor, and ceiling) of the accelerator room (5.2 m width × 5.6 m length × 2.7 m height, same as the experimental site). The 5.2 m-wide concrete wall was contacted with the detector probe without Pb shield in this calculation. Here, it was assumed that the ^{152}Eu and ^{60}Co point sources were arranged three-dimensionally at 1 cm steps in the crosswise direction and in 0.5 cm steps in the depth direction up to 50 cm, and distributed uniformly in the six concrete bodies. Specially, in the 30 cm × 30 cm × 20 cm depth area in contact with the detector probe, the sources were arranged at 0.5 cm steps in the crosswise direction. The dose rate from each source was calculated by the following formula

$$\text{Dose rate} = A \times F_a / r^2,$$

where A is the ^{152}Eu or ^{60}Co activity, F_a is the effective dose transmission of γ-rays across 2.1 g/cm^3 concrete (taken from a reference[6]), and r is the distance between the source and detector. Because the concrete density in the experimental site was 2.3 g/cm^3, the calculated dose rates were corrected by multiplying by 2.1/2.3. The dose rate from an individual concrete face was calculated by adding all sources.

The estimated contribution ratios to the dose rate from the six concrete bodies are shown in Figure 4. Two contact places were considered: at the centre of the wall (260 cm from the right end of the wall and 135 cm above the floor) and near the corner of the wall (50 cm from the right end of the wall and 50 cm above the floor). The contribution ratio of the dose rate from the contact wall was found to be independent of the nuclide (^{152}Eu or ^{60}Co) and contact place (centre or corner), and it was estimated to be approximately 50%. Therefore, the ambient dose rate must be suppressed by a Pb shield for accurate dose rate measurement.

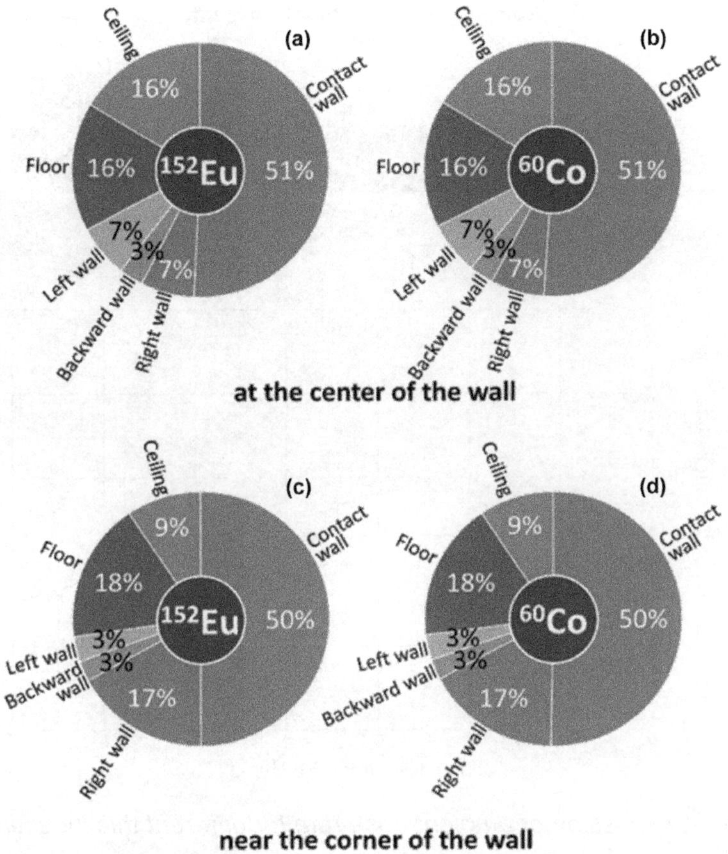

Figure 4 *Estimated contribution ratios of dose rate from six concrete faces of the accelerator room: at the centre of the wall, from (a)* ^{152}*Eu and (b)* ^{60}*Co, and near the corner of the wall, from (c)* ^{152}*Eu and (d)* ^{60}*Co. The activity was assumed to be distributed uniformly in all the concrete (see the text for further details).*

3.2 Minimisation of Pb shield for the survey meter

To determine the ambient dose rate transmission through each Pb shield, the dose rate was measured on a 10 cm thick Pb block placed on the walls and floor of the accelerator room at 18 locations, as shown at the top of Figure 5. The Pb block could completely shield the dose rate from the contact concrete. For the different Pb shields, the average ambient dose rate transmissions across them are plotted in the bottom of Figure 5. The transmission decreased exponentially with the thickness of the Pb shield.

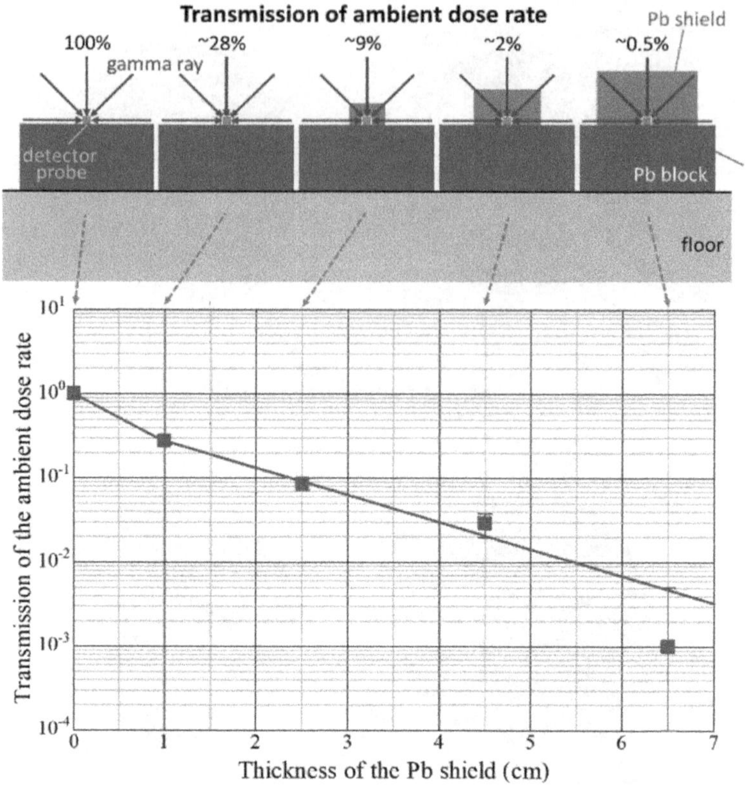

Figure 5 *Transmission of ambient dose rate for different thicknesses of the Pb shield.*

As mentioned in the previous section, approximately 50% of the dose rate comes from sources other than the contacted concrete, when the detector probe was used without the Pb shield. Therefore, the contribution ratios of ambient dose rate were estimated to be approximately 50, 14, 4, 1, and 0.3% for 0, 1, 2.5, 4.5, and 6.5 cm thick Pb shields, respectively. The 1% contribution for the 4.5 cm thick Pb shield is small enough for dose rate measurement, while a thicker Pb shield did not significantly improve the accuracy. The weight of the 4.5 cm Pb shield for the customised T-SP2 was 16 kg. When divided into two bodies, it was light enough to be carried manually.

3.3 Effect of Pb shield thickness on measured dose rate

The contact dose rate as a function of the specific activity of ^{152}Eu + ^{60}Co on the walls and floor of the accelerator room using Pb shields of 0, 1.0, 2.5, 4.5, and 6.5 cm thickness are shown in Figure 6. Here, the contact dose rates are plotted as "net contact dose rate" by subtracting the background (0.17, 0.09, 0.07, 0.06, and 0.06 μSv/h for 0, 1.0, 2.5, 4.5, and 6.5 cm thick Pb shields, respectively), which were measured in the next non-activated room.

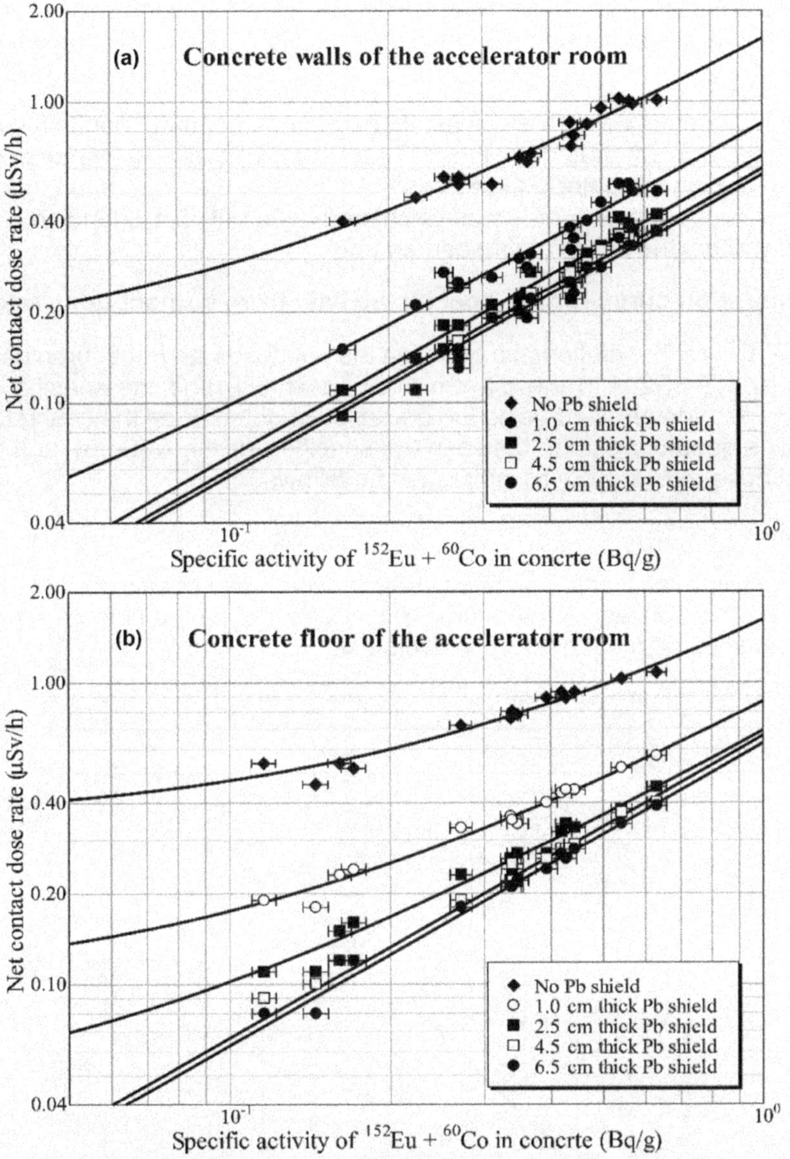

Figure 6 *Variations of the contact dose rates as a function of the specific activity of ^{152}Eu + ^{60}Co (a) on the walls and (b) on the floor of the accelerator room for 0, 1.0, 2.5, 4.5, and 6.5 cm of Pb shield thickness.*

The net contact dose rate depends on the specific activity in the concrete. For a specific activity, the net contact dose rate increases as the Pb shield becomes thinner due to ambient dose rate transmitted across the Pb shield. The net contact dose rates were almost the same between the 4.5 and 6.5 cm thick Pb shields. This is consistent with the aforementioned result that the transmitted ambient dose rate becomes negligible when the Pb shield is 4.5 cm or thicker.

The net contact dose rates on the walls were higher than those on the floor. This result was attributed to the different densities of the reinforcing rods between the wall and floor. In the case of the floor, the dose rate contribution ratio from the reinforcing rods was calculated to be only 9.4%, and so it did not seriously affect the dose rate measurement.

3.4 Calibration curve for the specific activity from contact dose rate

Figure 7 shows the relationship between the net dose rate (measured using the customised T-SP2 with the 6.5 cm thick Pb shield) and the specific activity (^{152}Eu + ^{60}Co) determined using the portable Ge detector on the concrete floor. There is a good correlation between the specific activity, A (Bq/g), and the net contact dose rate, D (μSv/h) expressed as follows

$$A = 1.60 \times D.$$

Using this relationship as the calibration curve, the specific activity could be estimated immediately from the dose rate measured by the customised T-SP2 with a Pb shield 4.5 cm or more in thickness.

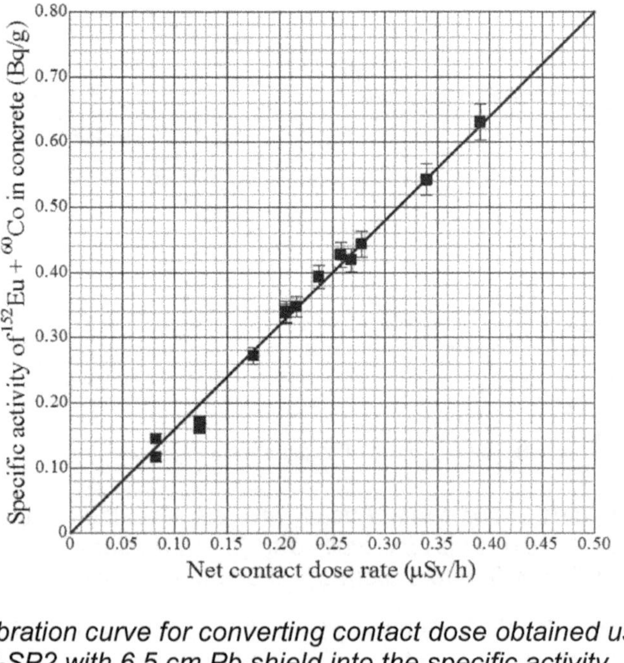

Figure 7 *Calibration curve for converting contact dose obtained using the customised T-SP2 with 6.5 cm Pb shield into the specific activity.*

3.5 Detection limit of specific activity using the survey meter

In the next room which was not activated, the background dose rates were measured repeatedly more than 40 times at the time constants of 3, 10, and 30 seconds. The detection limit was defined as 3σ, where σ was obtained from the standard deviation of these background dose rates. For both 4.5 and 6.5 cm thick Pb shields, the 3σ value was determined to be 0.04, 0.02, and 0.02 μSv/h at the time constant of 3, 10, and 30 seconds, respectively. In Japan, the clearance limits of ^{152}Eu and ^{60}Co in activated materials are both 0.1 Bq/g in specific activity. At a time constant of 3 seconds, the detection limit of 0.04 μSv/h corresponds to 0.06 Bq/g of the ^{152}Eu + ^{60}Co specific activity. That is, specific activities in the concrete at below the clearance limit could be measured this way.

3.6 Correction factor for slope of the calibration curve

The slope of the calibration curve obtained in this study is slightly affected by the $^{60}Co/^{152}Eu$ ratio in concrete, due to different effective dose equivalents between these two radionuclides. Although the specific activity ratio of $^{60}Co/^{152}Eu$ was determined to be average 0.28 in this experimental site, the ratio depends on machine operation duration and abundance ratio of stable Eu and Co in concrete.

Correction factor for the slope of the calibration curve was obtained by two steps. First, the dose rates from ^{152}Eu and ^{60}Co were calculated by a simplified method that was basically the same as that described above (in the subsection titled "Source of γ-rays measured as dose rate by the survey meter)". In this calculation, the detector probe was shielded by 6.5 cm thick Pb cylinder with a 7 cm diameter hole, and placed in contact with the centre of the concrete wall. Second, the correction factors for the specific activity ratio of $^{60}Co/^{152}Eu$ were derived from the relationship between the calculated dose rates from ^{152}Eu and ^{60}Co in the contact wall.

Figure 8 shows the variation of correction factor for the slope of calibration curve. The specific activity ratio of $^{60}Co/^{152}Eu$ is constant in all the concrete in the accelerator room, as obtained in previous work.[4] Therefore, the specific activities of ^{152}Eu and ^{60}Co only need to be determined at one place by using γ-ray spectrometer with high energy resolution, such as $CeBr_3$ scintillation detector and Ge semiconductor detector. Then, the calibration curve could be redrawn by using the slope correction factor obtained from the specific activity ratio of $^{60}Co/^{152}Eu$ in Figure 8.

4 CONCLUSIONS

An in situ method to determine the specific activity in concrete structures using a customised T-SP2 survey meter was established in this study. For dose rate measurement, a Pb shield of 4.5 cm thickness was enough to shield ambient γ-rays. The calibration curve for converting the net contact dose rate on concrete into the specific activity in concrete was obtained as A = 1.60 × D

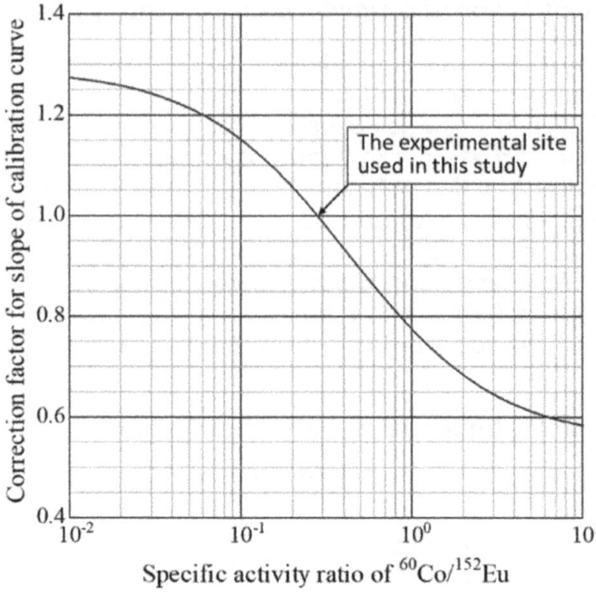

Figure 8 *Correction factor for the slope of the calibration curve.*

× C_f, where A, D, and C_f are the specific activity (Bq/g), the net contact dose rate (μSv/h), and the correction factor for the slope of the calibration curve, respectively. Further, this method could measure specific activity in the concrete at below the clearance limit in Japan. This method is able to be used as a roughly screening approach of in situ measurement.

CONFLICTS OF INTEREST

There are no conflicts to declare.

ACKNOWLEDGEMENTS

This work was supported by Radiation Safety Research Promotion Fund. This work was performed as one of Priority research areas for FY2017 and FY2018, Clearance of materials from the decommissioning of accelerator facilities, in Radiation Safety Research Program. The authors thank Drs. Toshioh Fujibuchi (Kyushu Univ.), Norihiro Matsuda (Japan Atomic Energy Agency), Shunsuke Yonai (National Institutes for Quantum and Radiological Science and Technology), Satoshi Nakamura (National Cancer Center Japan), Yoshitomo Uwamino (RIKEN), Koji Watabe (Tohoku University), and Minoru Ohkoshi (Japan Radioisotope Association) for their important advice to this study.

NOTES AND REFERNCES

1. K. Masumoto *et al.*, *Journal of Radioanalytical and Nuclear Chemistry*, 2003, **255**, 465.

2. K. Masumoto *et al.*, *Journal of Radioanalytical and Nuclear Chemistry*, 2008, **278**, 449.
3. N. Kinoshita *et al.*, *Nuclear Technology*, 2009, **168**, 694.
4. H. Matsumura *et al.*, *IOP Conf. Series: Journal of Physics: Conf. Series*, 2018, **1046**, 012016.
5. K. Kusumoto, K. Mizuno and R. Iwata, *Proceedings of the IVth International Workshop on Targetry and Target Chemistry (TARGETRY'91) in PSI-Proc. 92-01*, PSI Villigen, Switzerland, September, 1991.
6. Nuclear Safety Technology Center, *Data (radiation) Book for Shielding Calculation Task for Accelerator Facilities*, Nuclear Safety Technology Center, Tokyo, 2015, in Japanese.

TESTS OF NUCLEAR JET ENGINES AS A NEW SOURCE OF RADIOACTIVITY IN THE ENVIRONMENT

Jerzy W. Mietelski [a]

[a] Institute of Nuclear Physics, Polish Academy of Science, Radzikowskiego 152, 31-342 Kraków,

jerzy.mietelski@ifj.edu.pl

ABSTRACT

Russia announced that they successfully conducted test of nuclear jet engines in autumn of 2017. Construction details of the tested nuclear jet engine remain unknown, but the general concepts of such constructions are well known. The heat is generated by radioactive decay instead of combustion of liquid chemical fuel. There are two possible sources of radioactive heat: the very active radioisotope source (as in a Radioactive Thermoelectric Generator, RTG) or a nuclear reactor. Simple calculations reveal that the required activity for propulsion of cruise-class missiles is on the order of magnitude of an EBq (10^{18} Bq). It is likely, that the mysterious air contamination with $^{106}Ru/^{106}Rh$, observed in beginning of October 2017 all around Europe was a result of such tests. Tests of nuclear jet engines in open air is a new challenge for radioactivity monitoring systems in many countries. The use of RTG–like radioisotopes is a new situation for which many countries and international organisations are not prepared at all. Another important question is activity released during possible crash, which might happen during a test flight.

1 INTRODUCTION

Russia announced that they successfully conducted a test of nuclear jet engines in autumn of 2017.[1] Although the construction details of the tested nuclear jet engine(s) remain unknown, the general concept of such construction has been well known since the 1950s. The main advantage of such engines is the unlimited range of an aeroplane or any kind of mobile weapon driven by such an engine. The principle is that in place of the combustion chamber of any kind of jet engine (turbo jet, ram jet, turbo fan...) the heat is generated by radioactive decay instead of combustion of liquid chemical fuel.

2 ENERGETIC REQUIREMENTS

There are two possible sources of radioactive heat: a very active radioisotope source (e.g. a Radioactive Thermoelectric Generator, RTG) or a nuclear reactor.[2] In the case of an RTG-like isolated isotope source, ^{238}Pu, ^{244}Cm, $^{90}Sr/^{90}Y$ or $^{106}Ru/^{106}Rh$ can be used. The generated power available is a product of the radioactive source activity and the mean energy of a single decay of a given isotope. On the other hand, the power needed for propulsion of a plane is the product of the thrust of the engine and the speed of the plane. Since the announcement was about jet engines for propulsion of cruise missile the speed is about 200 m/s and thrust about 4.5 kN. This gives an energy level close to 1 MW. This estimation is very rough since the given above typical thrust of engine of cruise missile is a maximum thrust of jet engine which is produced only during the start. During flight the produced thrust is on level of about 30 %-40 % of the maximum. So the actual requirements for power are lower, since initially the nuclear jet engine missile can use classical auxiliary solid fuel rockets. On the other hand, there is also a limited efficiency of conversion of thermal energy of radioactive decays to useful work in the turbine. Let say that it is close to 50%. Those two effects nearly mutually compensate the each other. In general, the power produced in radioactive decay is a product of activity of source and mean energy released in single radioactive decay. Simple calculation reveals that if the mean energy of decay is on the level of 1 MeV the required activity (to obtain 1 MW output) is in the order of magnitude of EBq (10^{18} Bq). The exact value depends on energy available for the specific isotope decay used, the efficiency of self-absorption of radiation etc.

10^{18}Bq is the order of magnitude of release of radionuclides in the severest of nuclear reactor accidents. It means that to drive a cruise missile the activity on board of this projectile has to be almost on the level of activity released (not the inventory) during the Chernobyl accident. It is no matter if it is either a small nuclear reactor (~1 MW power) or relatively very strong RTG source.

3 HYPOTHESIS ON ENVIRONMENTAL IMPACTS

It is likely, that the mysterious air contamination with $^{106}Ru/^{106}Rh$, observed at the beginning of October 2017 all around Europe[3] was a result of such tests.[4] Since the release was estimated to be on level of 0.1 PBq,[4] it means that less than 1 ppm of on-board activity was released. This could be a result of diffusive processes inside the fuel element due to the very high temperature of it during flight. Similarly, the nuclear reactor used in a jet engine could likely also release in diffusive processes volatile fission and activation products, since due to weight restrictions for flying objects it is very unlikely that construction will have separate intermediate cooling circuits. Moreover, due to lack of proper shielding (the biological shields are not necessary for an unmanned machine) the cruise missile driven by a reactor will have activated all the metal parts of engine and possibly even the whole machine. It will produce also a trace activation of air. On the other hand, since a cruise-class missile is a single-use device, the burn-up in the reactor will be relatively short, so no accumulation of long lived radionuclides is expected. Therefore, during flight test of a nuclear jet engine driven by a flow-through nuclear reactor one can expect mostly short lived fission products released to the environment, namely those with half-lives in range of hours - not days or longer. For instance it will be ^{130}I, ^{132}I and ^{133}I and in lower amounts ^{131}I. Radiocaesium (especially ^{137}Cs) will be likely produced in very small amounts. It could be different if instead of a cruise missile one considers a global range recce-drone driven by nuclear propulsion. There, due to much longer exploitation, and in consequence longer burn-up of nuclear fuel, the build-up of long lived radionuclides will be more pronounced. Nevertheless, the possible releases from any kind of nuclear reactor are a kind of releases to which all existing networks are more or less prepared.

However, the use of RTG–like radioisotopes is a new situation for which many countries and international organizations are not prepared at all. For example, if such test would be conducted not with nuclide emitting gamma radiation (like $^{106}Ru/^{106}Rh$) but with pure alpha emitters such as ^{210}Po or ^{242}Cm – who and how fast will they notice any traces in air? The other important question is activity released during a possible crash.

4 CONCLUSIONS

It is clear that the testing of nuclear jet engines creates a new situation for all existing radioactive monitoring networks. Until now all international (e.g. CTBTO) or national monitoring systems are based on gamma spectrometric systems. Nobody is prepared for the situation when a single isotope in the form of extreme high energy (RTG-like) can be released into the air producing contamination not detectable by gamma ray spectrometry. The overlooking of such contamination is possible. On the contrary, in case of use of nuclear reactor the composition of activation products and fission products can be different from that typical for energetic reactor failure, however, it is very unlikely that it will be overlooked by existing networks working under routine procedures. The case of still mysterious contamination, which happened in October 2017 with $^{106}Ru/^{106}Rh$ in the air requires the start of a serious debate about need of modification of the existing monitoring networks equipment and procedures to face possibility of the presence of releases of separated radionuclide which are not gamma-emitters.

CONFLICTS OF INTEREST

There are no conflicts to declare.

REFERENCES

1. V. V. Putin Presidential Address to the Federal Assembly, http://en.kremlin.ru/events/president/news/56957, 2018
2. R. A. Gabrielli and G. Herdich, *Progress in Aerospace Sciences*, 2015, **79**, 92–113.
3. IRSN, http://www.irsn.fr/EN/newsroom/News/Documents/IRSN_Information-Report_Ruthenium-106-in-europe_20171109.pdf
4. J. W. Mietelski and P. Povinec, 2019, in preparation.

INVESTIGATION OF NEUTRON-FLUENCE MEASUREMENT METHODS FOR ESTIMATING NEUTRON-INDUCED ACTIVITY FROM AN ELECTROSTATIC ACCELERATOR SOURCE

Hajime Nakamura*[a], Hiroshi Matsumura[a], Go Yoshida[a], Akihiro Toyoda[a], Kazuyoshi Masumoto[a], Taichi Miura[a], Kimikazu Sasa[b] and Tetsuaki Moriguchi[b]

[a] *High Energy Accelerator Organization (KEK) 1-1, Oho, Tsukuba, Ibaraki* 305-0801, Japan. [b] University of Tsukuba, 1-1-1 Tennodai, Tsukuba, Ibaraki 305-8577, Japan.

hajime.nakamura@kek.jp

ABSTRACT

Three methods to detect neutrons emitted from an electrostatic accelerator were examined. These were gold foil activation, thermoluminescent dosimeters (TLDs), and CR-39. The gold-foil activation method has previously been proven effective in the measurement of neutrons emitted from cyclotron sources. In the case of electrostatic accelerators, a greater amount of fast neutrons and a smaller amount of thermal neutrons are observed compared to a cyclotron accelerator. Therefore, the error in calculating the thermal neutron flux became unacceptably large. By comparison the CR-39 detectors were apparently unable to measure thermal neutrons well, because some tracks were also generated by fast neutrons causing a signal interference. In the case of TLDs, as the material was not sensitive to fast neutrons, it was not affected by them, and hence could adequately measure the thermal neutron flux. However, it was found that in some cases its measurements could be offset by the influence of high dose rate of gamma rays.

1 INTRODUCTION

During the operation of accelerators, accelerated particles strike various accelerator components such as slits, monitors, and the beam dump, causing activation of these materials and with arising secondary neutrons also activating the surrounding peripheral structures. The activity induced by neutrons is generally low, but the activated area can be large. In the case of decommissioning, it is very important to estimate the activation of the surrounding structural materials, including building materials. Therefore, the evaluation of neutron fluence measurements during accelerator operation is indispensable. In cyclotron facilities, gold foils have previously been used as activation detectors with good success. In the case of electrostatic-type accelerators, the particle energy is generally low and thermal neutron fluence is also very low. Accordingly, we attempted to use several different neutron-detection methods, such as activation detectors and dosimeters, in order to evaluate neutron fluence and contrast their relative features and performances. An experiment was performed at the 6-MV tandem electrostatic accelerator at the University of Tsukuba, using proton beams to impinge a metal target under several conditions to generate neutrons.

2 EXPERIMENTAL

2.1 Methods for neutron fluence

2.1.1 Gold foil activation.[1] This method utilizes paired sets of gold foils, one with and one without a cadmium cover exposed to the accelerator's generate neutron flux for a set period. Relative radioactivity (dose) was simultaneously measured using an imaging plate. Post exposure, the reference gold foils were measured with a HPGe gamma spectrometer and the cumulative neutron flux calculated from the measured gamma radioactivity.

2.1.2 CR-39 plastic nuclear track detectors.[2] When charged particles, such as protons and α particles, enter CR-39, minor radiation damage (trajectory) can be caused. However, as neutrons have no charge and do not directly make damage tracks in CR-39, the material was combined with an enshrouding proton radiator and an α converter. The proton radiator generates protons by injection of fast neutrons and the α converter generates α particles by injection of thermal neutrons. These neutron-induced secondary particles then generate small tracks in the CR39. The number of small tracks (pits) generated is then quantified by a step process of chemical treatment (etching) and counting under an optical microscope.

2.1.3 Thermoluminescent dosimeters (TLDs).[3] Panasonic UD813PQ4 thermoluminescent dosimeters were used for neutron detection. Again, pairs of TLDs were utilised; one bare TLD and another TLD covered with a cadmium sheet (0.5 mm in thickness). The TLD has two types of phosphor, one of which is sensitive to both thermal neutrons and gamma rays, while the other is sensitive to only gamma rays. The difference in the accumulated energy between the TLD pairs denotes the thermal neutron dose. The dose is converted to flux when taking into account the exposure time.

2.2 Facility

Figure 1 shows a layout of the 6-MV tandem electrostatic accelerator. The numbers in figure 1 indicate the locations of the detectors installed around the accelerator. Most of the detectors were set on the surface of the accelerator and some on the floor of the laboratory or suspended above the accelerator, as described in Table 1. The locations of the detectors around the target are shown in detail in Figure 2. The A3 beam line was used in the experiment. The target was set in a chamber in the A3 line and two types of metals were used for the setup: stainless steel (SUS304) as the material for the accelerator beam enclosure and tantalum as the material for the Faraday cup.

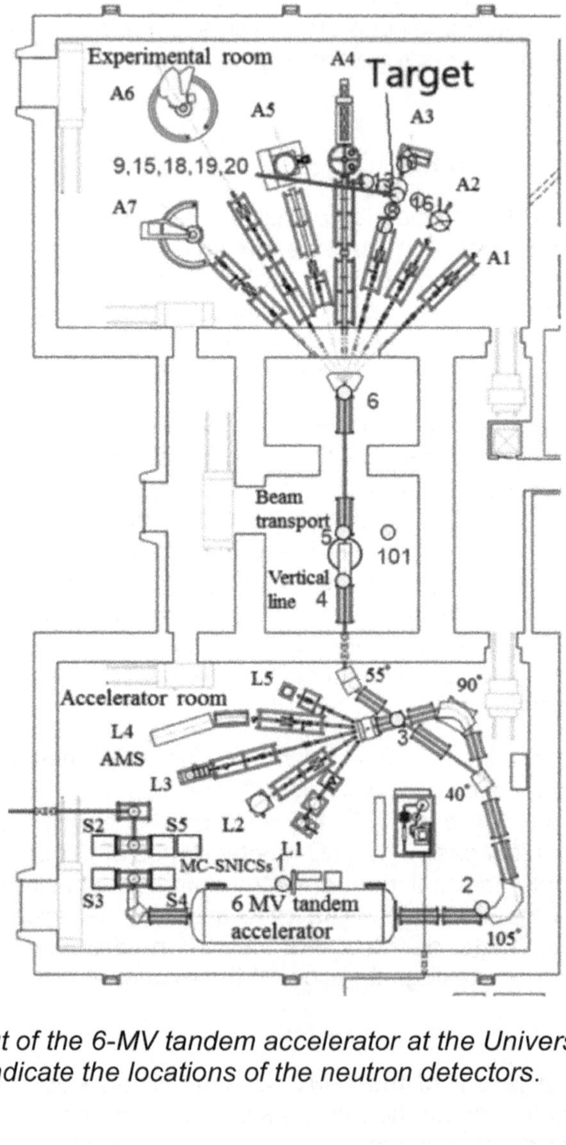

Figure 1 *Layout of the 6-MV tandem accelerator at the University of Tsukuba. The numbers indicate the locations of the neutron detectors.*

Table 1 *Detailed information on the detector's position. The facility has an Accelerator room, a beam transport room, and an experimental room. The detectors were installed on the surface of the accelerator, on the floor near the accelerator, above the target or aside of the target. Residual radiation dose before the first experiment and after the second experiment at each measurement point are also listed. (A): Accelerator room, (B): Beam transport room, (E): Experimental room, (-): no measurement.*

No.	Room	Condition	Detail	Residual radiation (μSv/h)	
				Before EXP.1st	After EXP.2nd
1	A	On Accelerator	Side of the accelerator tank	0.06	0.08
2	A	On Accelerator	Upstream of the 105° bending magnet	6.60	6.73
3	A	On Accelerator	Beam duct bellows downstream of the 90° bending magnet	0.95	1.17
4	B	On Accelerator	Upstream of the Faraday cup	0.73	11.4
5	B	On Accelerator	Upstream of the quadrupole magnet	10.0	30.0
101	B	On Floor	Floor 120 cm away from No.5	-	-
6	B	On Accelerator	Upstream of the magnet for beam delivery	6.00	30.0
7	E	On Accelerator	2 m upstream of the target	0.06	0.22
8	E	On Accelerator	1 m upstream of the target	0.06	0.28
9	E	On Accelerator	Window of the target chamber	0.07	0.94
15	E	On Floor	Floor 30 cm away from downstream of the target	0.05	0.22
18	E	Above Target	18 cm above the window of the target chamber	-	0.52
19	E	Above Target	68 cm above the window of the target chamber	-	0.23
20	E	Above Target	118 cm above the window of the target chamber	-	-
10	E	On Accelerator	1 m downstream of the target	0.07	0.20
13	E	Beside Target	50 cm horizontally from the target	0.08	0.27
14	E	Beside Target	1 m horizontally from the target	0.08	0.21
16	E	On Floor	Floor 120 cm away from the target	0.08	0.27

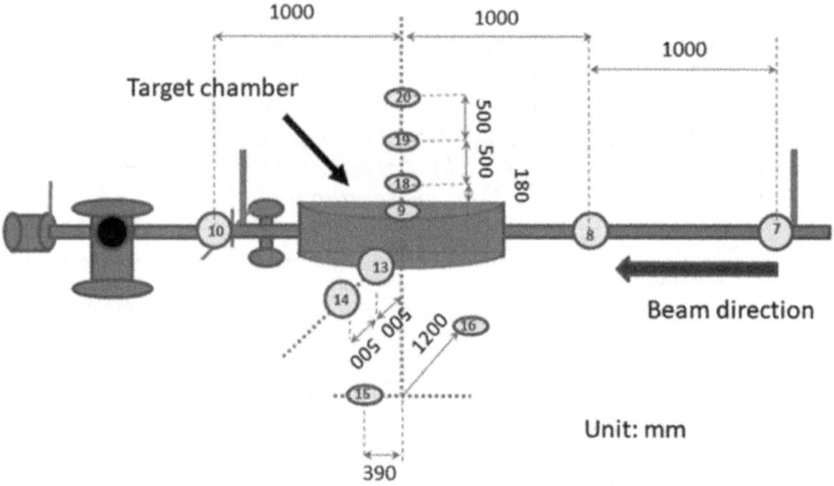

Figure 2 *Layout of the A3 beam line in the experimental room. The numbers indicate the locations of the detectors.*

The accelerator was operated in three modes. The first experiment utilised a flux of 12 MeV protons, at 1 µA current for a 2 h duration onto a stainless steel target. The second utilised the same conditions and exposure time but using a tantalum target. The final exposure utilised a 6 MeV proton beam at 1.3 µA, for an exposure of 2 h 15 m using a stainless steel target.

3 RESULTS AND DISCUSSION

3.1 Residual γ-dose determinations before and after experiment

The residual radiation doses at each measurement point before the first experiment and after the second experiment were measured using a NaI scintillation counter. The results are given in Table 1. The residual (background) radiation dose measured before the experiment at position No.2 in the accelerator room, and at No.5 and No.6 in the beam transport room were relatively high. At these locations, $^{54}Mn(T_{1/2}=312.2d)$, $^{56}Co(T_{1/2}=77.21d)$, and $^{57}Co(T_{1/2}=277.79d)$ were detected by follow-up measurements with a Ge detector. In the experimental room (hutch), the residual doses were determined to be at background levels.

The residual radiation dose measured after the second experiment in the accelerator room were not significantly higher than originally measured. However, a more notable increase for determined in the beam transport room and the experimental hutch. It was assumed that beam losses (leading to material activation) occurred mainly in the beam transport and the experimental room.

3.2 Determined neutron flux in the accelerator room and the beam transport room

Table 2 lists the determined thermal neutron flux ($cm^{-2}\cdot S^{-1}$) in the accelerator room and the beam transport measured using the CR-39 and TLDs. Because these locations were not affected by the target, detectors were installed during the 12 MeV operation period of the first experiment (SUS target) and the second experiment (Ta target) in succession. Gold foils were also installed; however, the activity of ^{198}Au could not be detected at the end of the experiment. Therefore, flux values are not listed in the table. In the third experiment (6 MeV SUS), gold foils were not installed. In the case of TLDs, thermal neutron at position No.6 was not measurable, because the gamma dose was 20±2.5mSv for the 2hrs exposure and the difference of dose between the two types of phosphor was within the gamma dose error. During the 12 MeV operation, the thermal neutron flux in the accelerator room was 10^1–10^2 $cm^{-2}\cdot S^{-1}$ measured with CR-39 and 10^0 to10^1 $cm^{-2}\cdot S^{-1}$ with TLDs. The thermal neutron flux in the beam transport room, measured with CR-39, was 10^2–10^4 $cm^{-2}\cdot S^{-1}$ and 10^1 to 10^2 $cm^{-2}\cdot S^{-1}$ with TLDs.

Table 2 *Thermal neutron flux ($cm^{-2}\cdot S^{-1}$) in the accelerator room and the beam transport measured with CR39 and TLDs. Because this place is not affected by the target, detectors were installed during the 12 MeV operation of the first experiment (SUS target) and the second experiment (Ta target).*

	EXP. 1st and 2nd 12 MeV SUS+Ta		EXP. 3rd 6MeV SUS	
No.	CR-39	TLD	CR-39	TLD
1	6.6E+01	(5.9±0.5) E+01	6.8E+01	(3.3±3.6) E+00
2	3.5E+02	(1.9±0.2)E+01	1.9E+01	(1.2±1.4)E+00
3	1.1E+01	(7.8±2.1)E+00	0.0E+00	(3.0±2.0)E+00
4	1.6E+03	(8.5±3.8)E+01	6.8E+01	(5.3±9.6) E+00
5	1.6E+04	(1.9±1.2)E+02	2.0E+04	(1.2±3.4)E+02
101	4.2E+02	(1.8±0.04)E+02	9.7E+00	(1.8±0.8)E+01
6	9.1E+04	Unavailable	1.9E+01	Unavailable

3.3 Neutron flux in the experimental room

Table 3 lists the thermal neutron flux in the experimental room, measured with CR-39, TLDs, and gold foils. The measurement with CR-39 at position No.9 in the first experiment could not be quantified because the neutron dose was so high that pit counting could not be performed. The measurements with TLDs at positions No.9 and 18 in the first experiment, No.9 in the second experiment, and Nos. 9, 13 and 18 in the third experiment were also unavailable. During the 12 MeV operation with the SUS target, the determined thermal neutron flux was 10^3–10^5 $cm^{-2}\cdot S^{-1}$ measured with CR-39, 10^2–10^3 $cm^{-2}\cdot S^{-1}$ with TLDs, and 10^1–10^2 $cm^{-2}\cdot S^{-1}$ with gold foils. During the 12 MeV operation with the Ta target, the thermal neutron flux measured with CR-39 was 10^3–10^4, 10^2 with TLDs, and 10^0–10^2 $cm^{-2}\cdot S^{-1}$ with gold foils. During the 6 MeV operation with the SUS target, the thermal neutron flux measured with CR-39 was 10^1–10^3 $cm^{-2}\cdot S^{-1}$ and 10^0–10^1 $cm^{-2}\cdot S^{-1}$ with TLDs. The gold foils were not installed.

Table 3 *Thermal neutron flux ($cm^{-2}\cdot S^{-1}$) in the Experimental room measured with CR-39, TLDs, and gold foils. The first experiment was operated at 12 MeV with SUS target. The second one was operated at 12 MeV with the Ta target. The third one was 6 MeV with the SUS target.*

	12 MeV SUS Exp. 1st				12 MeV Ta EXP. 2nd				6 MeV SUS EXP. 3rd	
No.	CR-39	TLD	Gold thermal	Gold epithermal	CR-39	TLD	Gold thermal	Gold epithermal	CR-39	TLD
7	4.3E+03	(5.4±0.4)E+02	(3.0±2.6)E+01	(1.8±0.2)E+00	1.8E+03	(1.7±0.5)E+02	Not detectable	(2.7±0.2)E+00	8.8E+01	(8.1±5.2)E+00
8	1.2E+04	(6.5±0.9)E+02	(9.2±7.9)E+01	(2.4±0.2)E+00	5.8E+03	(2.7±0.3)E+02	Not detectable	(1.6±0.1)E+00	1.6E+02	(8.6±3.9)E+00
9	Unavailable	Unavailable	(9.5±8.3)E+02	(1.3±0.1)E+02	3.5E+05	Unavailable	(5.3±4.5E)+02	(5.1±0.4)E+01	8.4E+03	Unavailable
15	6.5E+03	(4.0±0.5)E+02	(1.3±1.1)E+02	(1.1±0.1)E+01	4.3E+03	(1.6±0.2)E+02	Not installed	Not installed	2.9E+01	(7.9±3.8)E+00
18	1.5E+05	Unavailable	(2.3±2.0)E+02	(1.6±0.1E+01	9.6E+04	(1.5±1.1)E+02	(9.6±8.3)E+01	(6.7±0.6)E+00	3.7E+03	Unavailable
19	1.4E+04	(3.7±1.5)E+02	(1.4±1.2)E+02	(2.9±0.3)E+00	2.3E+04	(2.4±0.2)E+02	(1.4±1.2)E+01	(1.5±0.1)E+00	5.1E+02	(2.0±0.8)E+01
20	9.6E+03	(7.1±0.9)E+02	(6.0±5.1)E+01	(2.7±0.2)E+00	6.8E+03	(2.5±0.7)E+02	(8.5±9.3)E+00	(1.8±0.2)E+00	7.6E+02	(2.7±0.3)E+01
10	1.4E+04	(4.3±0.5)E+02	(7.5±6.5)E+01	(2.6±0.2)E+00	4.7E+03	(1.6±0.6)E+02	(1.0±0.9)E+01	(3.7±0.3)E-01	1.9E+02	(1.0±0.4)E+01
13	5.6E+04	(2.0±0.2)E+03	(4.8±4.1)E+02	(1.8±0.2)E+01	3.9E+04	(5.8±1.4)E+02	Not installed	Not installed	1.6E+03	Unavailable
14	1.6E+04	(3.5±0.5)E+02	(2.0±1.7)E+02	(3.5±0.3)E+00	7.3E+03	(2.5±0.4)E+02	(4.0±3.5)E+01	(1.6±0.1)E+00	5.3E+02	(9.2±6.8)E+00
16	1.1E+04	(8.3±2.1)E+02	(9.4±8.0)E+02	(1.3±0.1)E+01	6.4E+03	(4.7±0.9E+02	(1.8±1.6)E+02	(4.8±0.4)E+00	6.1E+02	(2.0±0.8)E+01

3.4 Comparison of the three methods

In these experiments, the neutron flux values determined by the CR-39 was notably higher than for the other methods. The thermal neutron dose measured by CR-39 at position No.16 for the 12 MeV operation with the SUS target was 0.6 mSv and the fast neutron dose was 21.9 mSv. The number of thermal neutrons was obtained by removing the number of the pits made by fast neutrons in CR-39 from the total number of pits (the pits can be discerned). It was assumed that there was interference with the measurement of thermal neutrons when the fast neutrons flux was larger than thermal neutron flux.

In the gold foil activation method, because the Cd ratio (radioactivity in bare/Cd cover) was nearly 1, it was difficult to calculate the thermal neutron flux calculated from the radioactivity difference between the bare and Cd covered gold foil accurately. Hence this method proved largely ineffective.

In the TLD method, TLDs can measure a small thermal neutron flux more reliably than gold foils because they are not influenced by fast neutrons and act as integrated dosimeters; recording the combined gamma and neutron dose without differentiation. However, this also meant that reliable neutron measurements could not be recorded in locations around the accelerator where there was a high gamma field, such as positions No.6 and No.9.

The values obtained were also compared with the Monte Carlo calculation (PHITS code) by private communication. The ratios of C/M (calculated value/measured value) at the measurement points were 1–40 with gold foils and 0.7–5 with TLDs.

4 CONCLUSIONS

Neutron flux in the radiation field generated by an electrostatic accelerator was measured via three different methods.

The gold foil activation method is usually the main method for measuring thermal neutron flux, and it is very effective in the radiation field of cyclotron accelerators where the flux is high. In the radiation field of the electrostatic accelerator, there were cases where gold foils were not sufficiently activated for accurate measurement of low neutron fluxes to be made. Because TLDs are integrated dosimeters, they are able to measure a much small neutron flux than achievable with gold foils, though they are not affected by fast neutrons. Thus, they present the advantage of offering a more precise measurement. However, when there is a coincident high-dose gamma ray field TLD measurement are problematic because they are affected by radiation types. By contrast, CR-39 detectors recorded higher neutron flux values than other methods. This was attributable to the observation that both thermal and fast neutrons are measured by the CR-39, even for a device without a secondary particle scatter-shroud.

CONFLICTS OF INTEREST

There are no conflicts of interest to declare.

ACKNOWLEDGEMENTS

This research was carried out with the support of Radiation Safety Research Promotion Fund by the Nuclear Regulation Authority in Japan. The measurement was helped by Takayuki Nakabayashi, Japan Environment Research Co., Ltd.

NOTES AND REFERENCES

1. K. Masumto, A. Toyoda, K. Eda and T. Ishihara, *Radiation Safety Management*, 2002, **1**, 12.
2. K. Oda, M. Ito, Y Miyake, J. Yamamoto and T. Turuta, *Nuclear Instruments and Methods*, 1991, **61**, 302.
3. H. Nakamura, T. Miura, M. Numajiri, N. Nakao, H. Nakashima, N. Matsuda and S. Watanabe, *Radiation Protection Dosimetry*, 2005, **116**, 252.

ASSESSMENT OF SR-90, CS-137, NATURAL RADIONUCLIDES AND METALS IN MARINE FISH SPECIES CONSUMED IN THE CITY OF SÃO PAULO – BRAZIL

Marcelo Bessa Nisti*[a], Cátia Heloisa Rosignoli Saueia[a], Bruna Castilho[a] and Barbara Paci Mazzilli[a]

[a] Instituto de Pesquisas Energéticas e Nucleares, São Paulo, Brazil.

mbnisti@ipen.br

ABSTRACT

The aim of this study is to quantify the concentration of the ^{90}Sr, ^{137}Cs, natural radionuclides (^{210}Pb, ^{226}Ra and ^{228}Ra) and metals (As, Co, Cs, Rb, Se, Th, U and Zn) in the flesh of marine fish species most consumed in São Paulo city. The radionuclides and metals concentration were low (but close to the level of detection) or below the MDA of gamma spectrometry, liquid scintillation counting and instrumental neutron activation analysis. The radionuclides and metals concentrations are below limit adopted by the Brazilian Regulatory Agency.

1 INTRODUCTION

Fish is an excellent nutritional quality food; therefore, in recent years consumption has significantly increased in Brazil. Fish also represents an important part of the alimentary diet. Contamination from radionuclides and metals in an ecosystem can be significant and can build up in food to a high degree by bioaccumulation. As a result of bioaccumulation radionuclides and metals in the food chain can also cause severe health damage from the consumption pathway.

Studying the dispersion of artificial radionuclides in the environment is important to determine the impact of nuclear waste discharges, nuclear accidents and nuclear weapons testing. The accident at the Fukushima Daiichi Nuclear Power Plant on 11 March 2011, released several radionuclides into the marine environment by aerial deposition and liquid discharge, with various levels of radioactivity. ^{90}Sr and ^{137}Cs were two of these elements released into the environment. The atmospheric releases of ^{90}Sr and ^{137}Cs are estimated to have been in the ranges from $3.3.10^{-3}$ to 0.14 PBq and from 7 to 20 PBq, respectively[1]. ^{90}Sr and ^{137}Cs are produced by nuclear fission with a physical half-life of 28.79 years and 30.15 years, respectively[2]. The natural radionuclides (^{210}Pb, ^{226}Ra and ^{228}Ra) were determined to compare with the literature values. The concentrations of metals were determined to check with the values of the maximum limits adopted by the Brazilian legislation[3,4].

The aim of this study is to quantify the concentration of the ^{90}Sr, ^{137}Cs, natural radionuclides and metals in the flesh of marine fish species most consumed in São Paulo city. The determination of ^{90}Sr by liquid scintillation counting (LSC) measurement was performed using a 1220 Quantulus™ Ultra Low Level Liquid Scintillation Spectrometer. ^{137}Cs and natural radionuclides were measured by gamma spectrometry, using an HPGe detector. The metal concentrations (As, Co, Cs, Rb, Se, Th, U and Zn) were determined by instrumental neutron activation analysis (INAA).

2 MATERIALS AND METHODS

The fish samples were purchased in a São Paulo city supermarket with variety of fish species of various origins. The fish species analysed were bluefish (*pomatomus saltatrix*), croaker (*argyrosomus regius*), hake (*merluccius merluccius*), smooth-hounds nei (*mustelus spp.*), sardine (*sardina pilchardus*) and tuna fish (*thunnus spp.*).

About 0.6 kg of flesh from the fish samples were calcined at 450 °C, resulting between 12 g to 28 g of ash depending on the species.

The resultant fish ash was packed in a polyethylene flask and sealed for about four weeks prior to the measurement, in order to ensure that equilibrium had been reached (between ^{226}Ra and its decay products of short half-life). The fish samples were measured (^{210}Pb, ^{226}Ra, ^{228}Ra and ^{137}Cs) by gamma spectrometry with two Hyper Pure Germanium detector HPGe, GX4020 with 47% relative efficiency and GX2518 with 25% relative efficiency, from CAMBERRA with associated electronics and coupled to a microcomputer. The counting time used 400,000 - 600,000 s[5]. The radionuclide ^{226}Ra was determined taking into account the mean value of the photo peaks of the radionuclides: ^{214}Pb (295.21 keV and 351.92 keV) and ^{214}Bi (609.31 keV and 1120.30 keV). The radionuclide ^{228}Ra was determined by measuring the intensity of the photo peaks 911.07 keV and 969.11 keV from ^{228}Ac. The radionuclide ^{210}Pb was measured through its photo peak of 46.50 keV[6]. Multichannel Maestro was used for spectrum acquisition[7]. The gamma spectra were analysed with the InterWinner 6.0 from Eurisys Measures[8], taking into account the gamma peak area, the background radiation for the gamma transition considered, the sample weight, the detector efficiency and the counting time. The determination of the minimum detectable activity (MDA) was made by the model proposed by Currie[9].

The determination of ^{90}Sr was performed by liquid scintillation counting (LSC) using a 1220 Quantulus™ Ultra Low Level Liquid Scintillation Spectrometer[10]. About 6 g to 12 g of the fish ash was digested on a microwave MARS with 1000 µL of stable Sr carrier. The sample was evaporated and dissolved in 8M HNO_3. The solution was loaded onto the Sr-resin column preconditioned in advance with 8M HNO_3. ^{90}Sr was retained and ^{90}Y was eluted at a flow rate 6 drops min^{-1}. ^{90}Sr was eluted with 30 mL of 0.5 M HNO_3, 0.3 g of oxalic acid was added with drops of NH_3 under gently shaking and strontium was precipitated. The precipitate was weighed to calculate the chemical recovery by gravimetric method. The precipitate was transferred quantitatively into an appropriate vial with the paper filter 1 mL of 1M HNO_3 was added to dissolve the precipitate. Finally, 15 mL of scintillation solution UltimaGold AB was added to the vial and the final solution was counted[11]. The chemical yield ranged from 60% to 78%. The channel range used was from 300 to 700. The counting time used 86,000 s.

The metal concentrations were determined by instrumental neutron activation analysis (INAA). The determination of the metals (As, Co, Cs, Rb, Se, Th, U and Zn) was carried out by irradiation of approximately 150 mg of ash, during 6 hours at a neutron flux of 10^{12} $n.cm^{-2}s^{-1}$, at the nuclear research reactor IEA-R1 of Instituto de Pesquisas Energéticas e Nucleares (IPEN). The induced radioactivity was measured by gamma-spectrometry with a hyper-pure germanium detector, EGPC 25 coaxial, with 25% relative efficiency and resolution of 2.10 keV. The concentration of the analysed elements was determined by comparing activities obtained in the samples with standard reference materials IAEA SL-1 Lake Sediment, IAEA SL-3 Lake Sediment and NIST- Montana Soil. The spectra analyses were made by WinnerGamma program on InterWinner[6]. The determination of the Minimum Detectable Activity (MDA) was made by the model proposed by Keith *et al*[12].

3 RESULTS AND DISCUSSION

The results of the ^{137}Cs, ^{210}Pb, ^{226}Ra and ^{228}Ra (Bq kg^{-1}, wet weight) in six species of fish are presented in Table 1, together with the world interval of UNSCEAR[13].

The ^{210}Pb, ^{226}Ra, ^{228}Ra and ^{137}Cs concentration were low (but close to the level of detection) or below the MDA of gamma spectrometry. The MDA "*a posteriori*" were different due to the mass variation used for each fish sample. The activity concentration of ^{210}Pb, ^{226}Ra and ^{228}Ra in the fish are of the same order of magnitude of world interval values of UNSCEAR[13].

Table 1 *^{137}Cs, ^{210}Pb, ^{226}Ra and ^{228}Ra (Bq kg^{-1}, wet weight) in six species of fish.*

Fish sample	Cs-137	Pb-210	Ra-226	Ra-228
Bluefish	0.11±0,02	< 0.76	< 0.19	< 0.32
Tuna fish	0.13±0.04	< 0.96	0.15±0.05	0.31±0.12
Smooth-hounds nei	< 0.19	< 1.3	0.23±0.08	0.37±0.18
Croaker	< 0.33	< 2.2	0.58±0.14	0.71±0.30
Hake	< 0.08	< 0.63	< 0.17	< 0.31
Sardine	< 0.11	< 0.95	0.24±0.08	0.27±0.09
Interval (UNSCEAR)[13]	-	0.014 – 4.8	0.0085 – 7.4	0.32

The ^{90}Sr concentration was below the MDA (< 0.02 Bq kg^{-1}) of liquid scintillation counting for all fish samples. Godoy *et al*[14] determined activity concentration of ^{137}Cs and ^{90}Sr in fish from Brazilian southeastern coastal region, with a range 0.03 to 1.7 Bq kg^{-1} for ^{137}Cs and for ^{90}Sr with a range of < 0.006 to 0.18 Bq kg^{-1}. Cunha *et al*[15] analyzed levels of ^{137}Cs in fish from the Brazilian coast, levels varied from 0.1 to 0.3 Bq kg^{-1}. The artificial radionuclides concentration obtained are in good agreement with values reported in literature. The artificial radionuclides ^{90}Sr and ^{137}Cs in the fish flesh samples studied are below limit adopted by the Brazilian regulatory guide (Comissão Nacional de Energia Nuclear- CNEN)[16].

The metals studied presented the MDA (μg g^{-1}, dry weight): As (<0.42 μg g^{-1}), Co (<0.75 μg g^{-1}), Cs (<0.51 μg g^{-1}), Rb (<5.9 μg g^{-1}), Se (< 0.40 μg g^{-1}), Th (<0.35 μg g^{-1}), U (<0.54 μg g^{-1}) and Zn (< 9 μg g^{-1}) of INAA. The results of the metals (As, Cs, Rb and Zn) (μg g^{-1}, wet weight) in the six species of fish are presented in Table 2, with comparison with Brazilian legislations[3,4].

The following elements presented concentration below the MDA for all fish samples: Co, Se, Th and U. The As and Zn concentrations are below limit adopted by the Brazilian legislation[3,4]. Fabiano *et al*[17] determined essential and toxic elements in some of the most consumed fish species in São Paulo city, with a range 0.02 to 4.1 mg kg^{-1} for As; 0.003 to 0.026 mg kg^{-1} for Cs; 0.07 to 1.5 mg kg^{-1} for Rb and 2.5 to 9.0 mg kg^{-1} for Zn. The results obtained for the elements (As, Cs, Rb and Zn) are in good agreement with literature values.

Table 2 *As, Cs, Rb and Zn (μg g^{-1}, wet weight) in six species of fish.*

Fish samples	Weight ratio dry/wet	As	Cs	Rb	Zn
Bluefish	0.037	nd	0.05±0.01	0.79±0.03	9.5±0.3
Tuna fish	0.046	nd	0.03±0.01	0.59±0.05	5.9±0.2
Smooth-hounds nei	0.020	nd	0.03±0.01	1.0±0.1	nd
Croaker	0.041	0.21±0.02	nd	nd	nd
Hake	0.036	nd	nd	0.59±0.04	0.66±0.04
Sardine	0.043	nd	nd	0.99±0.09	2.2±0.1
Brazilian legislation[3,4]		1.0	-	-	50

nd: not detected.

The performance of the gamma spectrometry and LSC methods was evaluated by participating twice a year in Proficiency Tests (PT) organized by Instituto de Radioproteção e Dosimetria IRD/CNEN[18], indicating good performance of laboratory. Relative standard deviation and relative error were lower than 10% proving the precision and accuracy of the INAA technique.

4 CONCLUSIONS

The radionuclides (^{210}Pb, ^{226}Ra, ^{228}Ra, ^{137}Cs and ^{90}Sr) and metals (As, Co, Cs, Rb, Se, Th, U and Zn) concentration were low (but close to the level of detection) or below the MDA of gamma spectrometry, liquid scintillation counting and instrumental neutron activation analysis.

The results obtained for radionuclides and metals are in good agreement with literature values. The radionuclides and metals concentrations are below the limit adopted by the Brazilian Regulatory Agency. To ensure the quality of the marine fish consumed by the population, continuous monitoring is necessary to check a possible contamination.

The results obtained in this study can be used to verify the quality of the fish consumed in the city of São Paulo.

CONFLICTS OF INTEREST

There are no conflicts to declare.

REFERENCES

1. IAEA - International Atomic Energy Agency. The Fukushima Daiichi accident - Technical vol. 4 - Radiological consequences, ed. Vienna: International Atomic Energy Agency, 2015.
2. M. L´Annunziata. Handbook of Radioactivity Analysis, 3 edn, 2012.
3. Ministério da Saúde – MS - Agência Nacional de Vigilância Sanitária – ANVISA - Resolução da Diretoria Colegiada – RDC nº 42, de 29 de agosto de 2013.
4. BRASIL – Presidência da República - Decreto nº 55.871, de 26 de março de 1965.
5. M. B. Nisti, A. J. G. Santos, B. R. S. Pecequilo, M. F. Máduar, M. M. Alencar and S. R. D. Moreira, *J. Radioanal. Nucl. Chem.*, 2009, **281**, 283.
6. G. F. Knoll, *Radiation Detection and Measurement*, 3rd edn, John Wiley & Sons, Inc., New York, 1999.
7. Maestro Software: Maestro for Windows (Emulation software for Gamma Spectroscopy) Model A65-B1Version 3.04, EG&G ORTEC, 1995.
8. INTERWINNER. InterWinner (WinnerGamma) Spectroscopy Program Family Version 6.0, 2004.
9. L .A. Currie, *Anal. Chem.*, 1968, **40**, 586.
10. Quantulus, Instrument Manual-Wallac 1220 QuantulusTM Ultra Low Level Liquid Scintillation Spectrometer, PerkinElmer, 2009.
11. IAEA - Analytical Quality in Nuclear Applications No. IAEA/AQ/27. Rapid simultaneous determination of ^{89}sr and ^{90}Sr in milk: A procedure using Cerenkov and scintillation counting, ed. Vienna: International Atomic Energy Agency, 2013.
12. L. H. Keith, W. Crummet, J. P. Deegar, R. A. Libby, J. K. Taylor and G. Wenter, *Analytical Chemistry*, 1983, **55**(14), 2210.
13. UNSCEAR - UNITED NATIONS SCIENTIFIC COMMITTEE ON THE EFFECTS OF ATOMIC RADIATION. The 2000 Report to the General Assembly with scientific Annexes (Annex B). New York, 2000.
14. J. M. Godoy, Z. L. Carvalho, F. C. Fernandes , O. M. Danelon, A. C. M. Ferreira and L. A. Roldão, *J. Environ. Radioact.,* 2003, **70**, 193.
15. I. I. L. Cunha, C. S. Munita, P. P. Paiva and A. Teixeira, *Sci. Total Environ.*, 1993, **139/140**, 431–435.
16. Comissão Nacional de Energia Nuclear - CNEN – Diretrizes Básicas de Proteção Radiológica CNEN-NN-3.01 – Posição Regulatória 3.01/006, 2005.
17. K. C. Fabiano, A. P. S. Lima, M. B .A. Vasconcellos and E. G. Moreira, *J. Radioanal. Nucl. Chem.*, 2016, **309**, 383.
18. L. Tauhata, M. E. C. Vianna, A. E. Oliveira, A. C. M. Ferreira and C. C. S. Conceição, *Appl. Radiat. Isot.*, 2000, **56**, 409.

ACHIEVING THE END STATE OF A NUCLEAR LICENSED SITE

Mustafa Sajih*, Slimane Doudou, Emily Phipps and Steve Wickham

Galson Sciences Ltd, 5 Grosvenor House, Melton Road, Oakham, Rutland LE15 6AX, UK.

* ms@galson-sciences.co.uk

ABSTRACT

End State refers to the condition of a nuclear licensed site (or part of the site) following all physical decommissioning and clean-up activities required to make the site (or part of the site) available for its next use. To reach the End State, a site must be compliant with the requirments for release of the site from environmental regulations. Site operators may apply to the environment agencies for agreement for a site to be released from radioactive substances regulation (RSR). The release of the site will only be granted when all disposals of radioactive waste have ended and the site is in a state that will ensure a satisfactory standard of protection for people and the environment. A guidance document, known as "the GRR", on the release of nuclear sites from RSR was published in July 2018. A site specific assessement of the benefits and detriments of clean-up and radioactive waste disposal options must be carried out as a requirement of the GRR. This involves identification of potential disposal options, assesssment and comparison of options, with stakeholder engagement, and optimisation of the preferred option if any radioactive material is to be left on site. As part of the assessment of waste disposal options, the GRR requires that operators assess potential consequnces of inadvertent human intrusion into the site, and aqueous releases from features on the site after the End State has been reached. Calculations of potential impacts from human intrusion and aqueous release scenarios are carried out and compared against guidance levels specified in the GRR. These assessments should also show that the consequential effects on non-human species are acceptable. Results of these assessments allow consideration of the implications of leaving radioactivity associated with particular features and components of the site, and inform comparison of alternative End State options.

1 INTRODUCTION

In July 2018, the Environment Agency (EA), Natural Resources Wales (NRW), and the Scottish Environment Protection Agency (SEPA) (the environment agencies) published their guidance on requirements for release of nuclear sites from radioactive substances regulations (RSR). This guidance, known as the GRR[1], is for operators of nuclear sites that hold or intend to hold an environmental permit for the disposal of radioactive waste. It describes what operators need to do, over the lifetime of their site, so the site can be released from RSR when all activities involving the management of radioactive waste have ended, and the site reaches the End State.

The site End State describes the condition to which the site (land, structure and infrastructure) will be taken and, where necessary, should be accompanied by a description of the controls required to protect people and the environment from any residual hazards[2].

The GRR introduces the concept of a site reference state (SRS), defined as follows: the condition of a nuclear site when it is fully compliant with the requirements for release of the site from RSR. This condition may be achieved after an operator has completed all planned work involving radioactive substances, or after a subsequent period of control for the purpose of radiological protection.

It is important to note that achieving the End State may not necessarily immediately result in achieving the SRS (suitable for release from RSR), because there may be a need for a period of control for the purpose of radiological protection that extends beyond achieving the End State.

The Nuclear Decommissioning Authority (NDA), a non-departmental public body that owns 17 nuclear sites across England, Wales and Scotland, is responsible to Government for defining and delivering End States for its sites. The ultimate goal for their mission is to achieve the End State at all sites by 2125.

1.1 Aims of the GRR

The guidance sets out: (1) the requirement for optimised plans for the management of the radioactive wastes from decommissioning and clean-up of nuclear sites (2) the standards that must be met if those optimised plans identify that radioactive wastes are best managed by on-site disposal (3) the standards that a nuclear site must meet to enable it to be released from RSR.

1.2 GRR requirements

Site operators are required to address the GRR requirments[1] for their site to be released from RSR. Among other requirements, the guidance requires operators to:

Produce and develop a waste management plan (WMP) that sets out how all radioactive waste that will arise over the whole remaining life cycle of the site will be dealt with, and justify that the WMP is optimised (Requirement R2).

Produce and develop a site wide environmental safety case (SWESC) that demonstrates that people and the environment will be adequately protected from the radiological hazard and any non- radiological hazard associated with the radioactive substances remaining on the site, before and after the site is released from RSR (Requirement R7).

1.3 Preparing nuclear sites for release from regulation throughout their lifecycle

The lifecycle of a nuclear site (Figure 1) extends from its design and construction through operation to decommissioning and final clean-up of the site. Activities involving radioactive substances are regulated throughout this lifecycle to ensure that disposals of radioactive waste are managed safely, and the public and the environment are properly protected when the site is released from regulation.

The journey to release from RSR requires that choices are made about the optimum way of dealing with the radioactivity that remains on site after operations have ended. This includes consideration of whether radioactive wastes could be disposed on-site or should be removed from the site, taking account of the radiological impact on people and the local environment.

1.4 Release from RSR

Operators may apply to the environment agencies for release from RSR for part or all of a nuclear site. The regulator will only agree to release a site from RSR when the operator has demonstrated that:

- All disposals of radioactive waste on or from the site have ceased; and
- The SRS has been achieved (the condition of the site being fully compliant with the requirements for release from RSR)

RSR activities will have ceased when all activities on-site involving the production, receipt of disposal of radioactive waste are complete, and there is no waste on-site awaiting disposal. The SRS is achieved when the SWESC shows that the residual risks presented by radioactive substances remaining on the site are, and will continue to be, consistent with GRR dose guidance levels without the need for a period of control (Figure 2a).

If the WMP and SWESC rely on a period of control for the purpose of radiological protection of people and the environment, they must demonstrate that the proposed controls will deliver the required protection, and that the arrangements for applying controls can be maintained for the duration of the period of control (Figure 2b). In most cases, it is anticipated that a period of validation monitoring will be required to confirm that the state of the site is in accordance with the assumptions of the SWESC, before release from RSR.

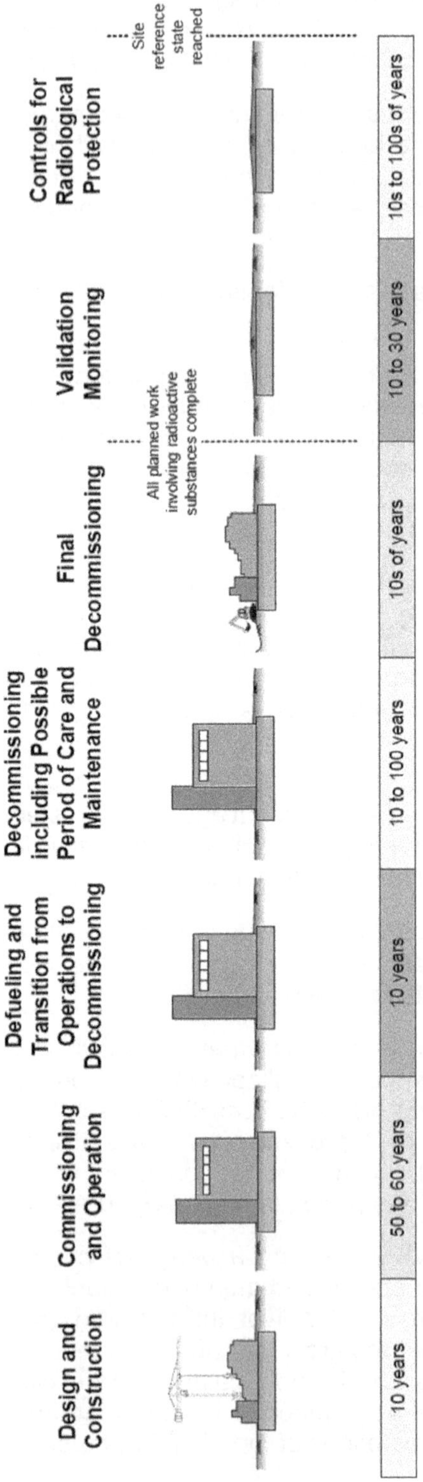

Figure 1 *Typical timeline for a nuclear site (from the GRR).*

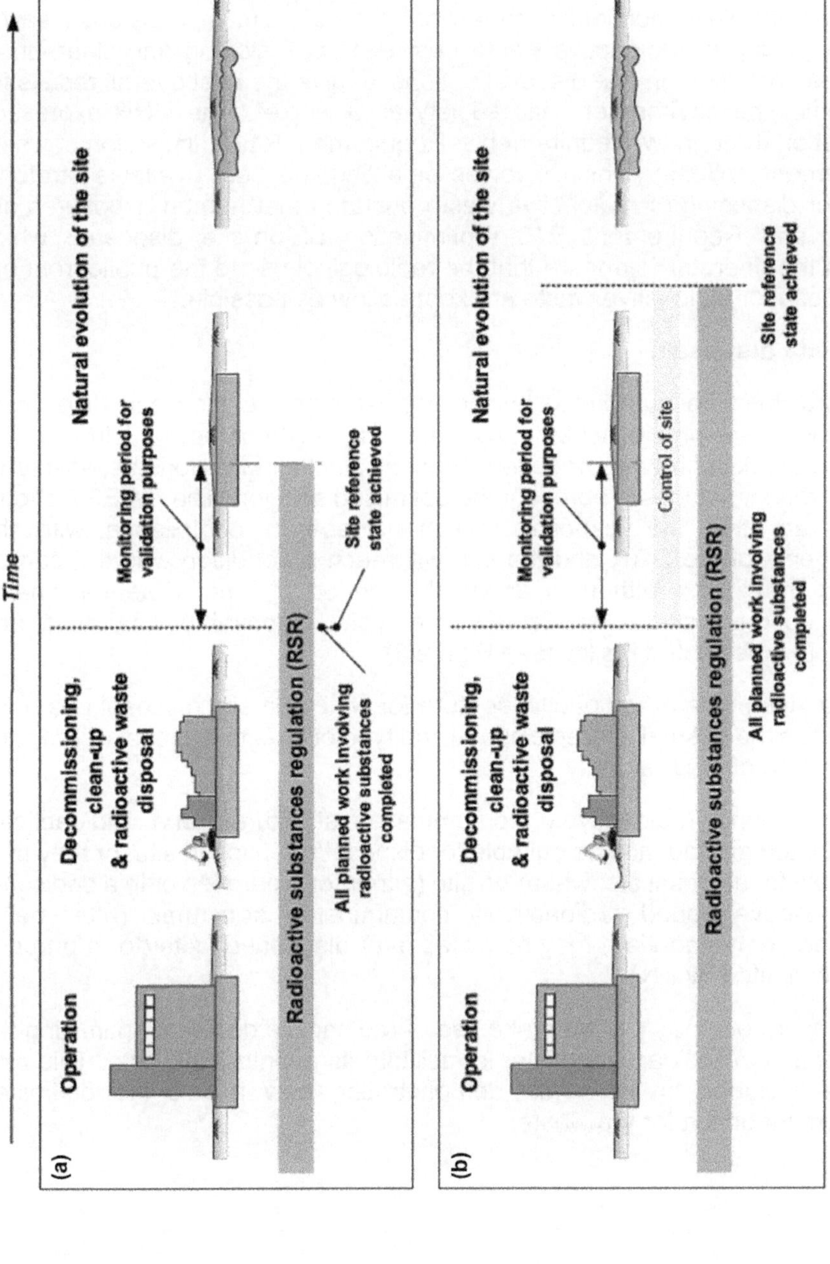

Figure 2 *Pathways to achieving the site reference state and release from RSR (from the GRR).*

2 WASTE MANAGEMENT OPTIONS AND OPTIMISATION OF DISPOSAL

The GRR requires operators of nuclear sites to ensure that exposures of the public to ionising radiation is kept below statutory limits, and that these exposures are kept as low as reasonably achievable (ALARA), taking into account economic and social factors. This is referred to as 'optimisation', an essential principle in radiological protection[3]. Operators must apply optimisation when disposing of radioactive waste from decommissioning and clean-up at their sites, including on-site disposal options, to give the best overall results for the people, the environment and society as a whole[4]. The GRR expresses optimisation through two requirements: Requirement R1 (optimisation of waste management options), which involves selecting the best available strategic option for disposing of radioactive waste ensuring that human exposures are ALARA, and Requirement R13 (optimisation of on-site disposal), which requires the operator to ensure that the radiological risk to the public from on-site disposal of radioactive waste are kept as low as possible.

2.1 On-site disposals

Operators must be authorised by the regulator before they can dispose of radioactive waste on-site, whether by emplacing waste or leaving it in situ. The WMP should describe the proposed disposals, including options assessments demonstrating that these represent the optimised solution. The SWESC should demonstrate that the disposals can be made in compliance with the requirements of the GRR and the site will reach a condition where it can be released from RSR within an acceptable period of time. Potential on-site disposal options include disposal of waste in situ, disposal for a purpose, and disposal in a dedicated facility (see Figure 3).

For a given type of radioactive feature for which on-site disposal has been identified as the optimal strategic option, the type of on-site disposal will depend on the nature of the feature.

Below-ground radioactively contaminated structures and radioactively contaminated ground may be suitable for disposal/retention in situ, or they may be suitable for disposal elsewhere on site (either for a purpose or in a dedicated facility). Above-ground radioactively contaminated structures (after being demolished or dismantled) may be suitable for disposal on site (for a purpose or in a dedicated facility).

Site operators may plan to make use of radioactive decay in optimising the management of radioactive waste, to facilitate its eventual disposal. This plan must be included in the WMP demonstrating how this is the optimised management option for the waste.

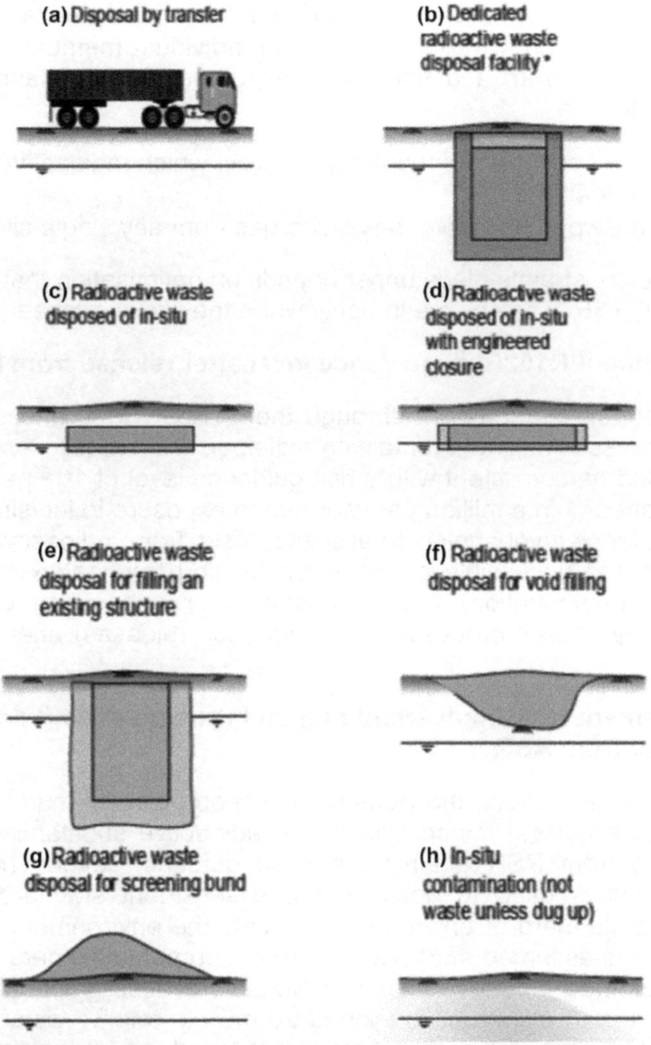

Figure 3 *Potential management options for radioactive substances from the final stages of decommissioning of nuclear sites (adapted from GRR).*

3 ADDRESSING THE RADIOLOGICAL REQUIREMENTS (REQUIREMENTS R9, R10 AND R11)

There are three quantitative radiological requirements in the GRR, with a dose constraint for the period prior to release from RSR, and two guidance levels, for natural evolution and human intrusion, for the period following release from RSR.

3.1 Requirement R9. Dose constraints during the period of RSR

During the period of RSR, the effective dose from the authorised site, to a representative person shall not exceed a source-related dose constraint and a site-related dose constraint. The environment agencies are required to have regard to the following maximum doses to individual members of the public which may result from a defined source, for use at the planning stage in radiation protection:

- 0.3 mSv per year from any source from which radioactive discharges are made; and
- 0.5 mSv per year from the discharges from any single site

The dose constraints place upper bounds on optimisation that apply during the period of RSR. They cease to apply when the site is released from RSR.

3.2 Requirement R10. Risk guidance level after release from RSR

Operators should demonstrate through the SWESC that, after release from RSR, the assessed risk from remaining radiological hazards to a representative person should be consistent with a risk guidance level of 10^{-6} per year (that is a risk of death of 1 in a million per year due to exposure to ionising radiation)[5]. The risk guidance level applies to assessed risks from radioactive substances dispersed in the accessible environment (arising from radioactive waste or radioactive contamination) due to migration or uncovering of radioactive substances by natural processes (e.g. aqueous releases) after release from RSR.

3.3 Requirement R10. Inadvertent human intrusion dose guidance level after release from RSR

Operators should assess the potential consequences of inadvertent human intrusion into any local concentrations of radioactive substances on the site after release from RSR against the dose guidance level. The term dose guidance level is used to describe the assessment standard, because it indicates the standard of environmental safety the environment agencies are looking for. The assumed effective dose to a representative person during and after the assumed intrusion should not exceed a dose guidance level in the range of around 3 mSv/year to around 20 mSv in total[6]. Values towards the lower end of this range are applicable to assessed exposures continuing over a period of years (prolonged exposures), while values towards the upper end of the range are applicable to assessed exposures that are only short term (transitory exposures).

Decisions to dispose of solid radioactive waste in the near-surface environment, or to allow radioactive contamination to remain in the ground, are made on the basis that it remains isolated from people until sufficient time has passed, to allow radioactive decay to reduce the risks to people and the environment to levels that are no longer radiologically significant. An inevitable consequence of such a decision is that, once a site has been released from RSR, local concentrations of radioactive substances remaining there may potentially be disturbed by people before sufficient time has passed. Such disturbance is termed “human intrusion”.

4 PERFORMANCE ASSESSMENT CALCULATIONS

As part of the assessment of waste disposal options, the GRR requires that operators assess potential consequences of inadvertent future human intrusion into the site and aqueous releases from features on the site after the site End State has been reached.

Calculations of potential impacts from human intrusion and aqueous release scenarios are carried out and compared against the guidance levels described above.

These assessments should also show that the consequential effects on non-human species are acceptable.

Results of these assessments help identify the implications of leaving radioactivity associated with particular components of the site and inform comparison of alternative End State options.

4.1 Inadvertent human intrusion

Inadvertent human intrusion encompasses a wide range of unintended actions that have the potential to disturb radioactive substances remaining on a site or to impair the ability to protect people and the environment, after knowledge of the location and nature of the site has been lost, in part or in full, to society[1]. Future human intrusion is assessed on the basis that it is likely to occur. Exposure to radioactivity as a result of human intrusion could occur at the time of intrusion, during processing or transport of excavated material, and after excavated material has been distributed.

Modelling considers a range of excavation sizes and potential scenarios for processing and use of excavated material (see Figure 4).

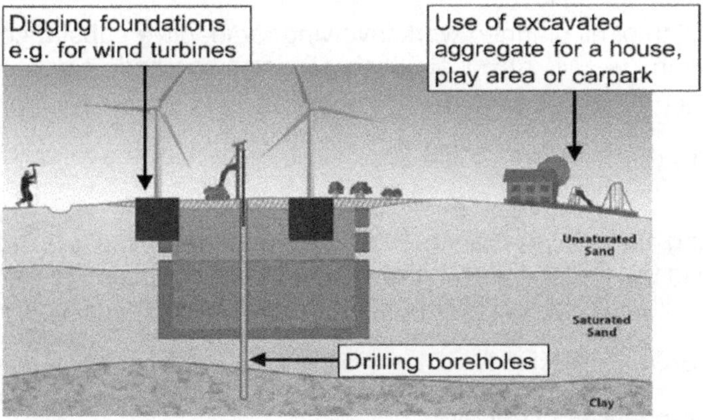

Figure 4 *Example of inadvertent human intrusion and subsequent use of excavated material.*

People undertaking excavation and post-excavation activities may be exposed to radioactivity through exposure pathways such as inhalation of dust, ingestion of dust and food, skin contamination and external exposure.

4.2 Aqueous releases

Modelling considers release of radionuclides from features left on the site and transport of these radionuclides through groundwater to areas where members of the public might be exposed in the future.

Potentially exposed groups could include campers, farmers, anglers and recreational users through exposure pathways such as:

- External exposure to contaminated ground or water
- Ingestion of fish or water
- Ingestion of plant or animal products
- Inadvertent ingestion of soil
- Inhalation of contaminated soils and dusts

5 CONCLUSIONS

For a nuclear site, the End State is not scheduled to be achieved for many decades, it is difficult to define the site End State in detail without ruling out credible options prematurely. There is an established decommissioning strategy for the UK nuclear sites which involves putting them into an inactive care and maintenance condition for a period of decades before completing the final stages of decommissioning. It is important to recognise the following potentially distinct milestones for a nuclear site (as a whole) during the final stages of decommissioning and clean-up:

1. Completion of reactor dismantling;

2. Completion of all planned work involving radioactive substances (including completion of all physical decommissioning and clean-up activities involving radioactive features of the site);

3. Achieving the site End State (the site can be made available for its next use);

4. Achieving the site reference state (the release of the site from RSR as defined in the GRR), following at least a period of validation monitoring and possibly a longer period of control for the purpose of radiological protection.

CONFLICTS OF INTEREST

There are no conflicts to declare.

REFERENCES

1. SEPA, EA, and NRW. Management of radioactive waste from decommissioning of nuclear sites: Guidance on Requirements for Release from Radioactive Substances Regulation Version 1.0, July 2018.
2. https://www.sepa.org.uk/media/365893/2018-07-17-grr-publication-v1-0.pdf
3. Nuclear Decommissioning Authority, Strategy, 2016. https://www.gov.uk/government/consultations/nuclear-decommissioning-authority-draft-strategy
4. Environment Agency, RSR: Principles of optimisation in the management and disposal of radioactive waste, 2010, Version 2.0. https://assets.publishing.service.gov.uk/government/uploads/system/uploads/attachment_data/file/296495/LIT_8452_a9c510.pdf
5. The environmental protection, England and Wales, the environmental permitting (England and Wales) regulations, 2016. http://www.legislation.gov.uk/uksi/2016/1154/pdfs/uksi_20161154_en.pdf.
6. HSE, Reducing Risks, Protecting People. HSE's decision-making process. HSE Books, Sheffield, 2001. http://www.hse.gov.uk/risk/theory/r2p2.pdf.
7. HPA, RCE-8 - Radiological Protection Objectives for the Land-based Disposal of Solid Radioactive Wastes: Advice from the Health Protection Agency, 2009. https://assets.publishing.service.gov.uk/government/uploads/system/uploads/attachment_data/file/335110/RCE-8_for_web_v2.pdf.

QUANTITATIVE EVALUATION OF RADIOACTIVITY IN CONCRETE AT PET CYCLOTRON FACILITY WITH SIMPLE AND NON-DESTRUCTIVE MEASUREMENT

Akihiro Toyoda*[a], Hiroshi Matsumura*[a], Kazuyoshi Masumoto*[a], Go Yoshida*[a], Taichi Miura*[a], Hajime Nakamura*[a], Kotaro Bessho*[a], Takayuki Nakabayashi*[b] and Genki Horitsugi*[c]

[a] High Energy Accelerator Organization (KEK) 1-1 Oho, Tsukuba, Ibaraki 305-0801, Japan, [b] Japan Environment Research. Co., Ltd., 6-24-1 Nish-Shinjyuku, Shinjyuku, Tokyo 160-0023, Japan, [c] Department of Nuclear Medicine and Tracer Kinetics Osaka University Graduate School of Medicine, 2-2 Yamadaoka, Suita, Osaka 565-0871, Japan

akihiro.toyoda@kek.jp

ABSTRACT

Accelerator operations often generate neutrons that may activate concrete in the accelerator room. When accelerators are decommissioned, the concrete will remain as radioactive waste. Not all of concrete in the room will be radioactive wastes, but at place with large amount of neutrons will be radioactive wastes. Therefore, it is very important to determine where and how many neutrons are generated during accelerator operation. We used gold foils to obtain this information in positron emission tomography (PET) cyclotron facilities. We obtained concrete core samples from a PET cyclotron to determine the amount of radioactivity contained in concrete. We obtained the operational parameters (particle, radioactive product, current, and operation time) of the PET cyclotron facility and estimated the radioactivity generated in concrete using the operational parameters and gold foil results. A comparison of the estimated and core sampling measurement results showed good agreement.

1 INTRODUCTION

In many small medical cyclotrons used for the production of positron emission tomography (PET) radiopharmaceuticals, neutrons generated by accelerator operations activate the concrete walls, floor, and ceiling. When decommissioning such cyclotrons, residual ^{152}Eu and ^{60}Co, both of which are long-lived isotopes and high-energy gamma-ray emitters, prove troublesome. Therefore, it is important to evaluate the radioactivity generated in the concrete structure.[1] At present, destructive sampling analysis is the only evaluation method available for this purpose. In this study, we developed a nondestructive method for analyzing the induced activity generated by accelerator operations. First, the flux of thermal and epithermal neutrons was determined using a gold foil activation detector. Then, residual radioactivity in concrete was estimated by using information on the cumulative operating record and neutron flux at the time. The estimated value was compared with the result obtained using the sampling method.

2 MEASUREMENT

2.1 Neutron flux monitoring during operation

We used a pair of gold foils (thickness: 20 μm, diameter: 6 mm) to measure the neutron flux in the PET cyclotron accelerator room during operation. One pair was bare, and the other was covered with Cd plates of 1 mm thickness and 10 mm square. In the cyclotron room, we pasted 10 or more gold foil pairs on the concrete walls, floor, and surface of the cyclotron before operation. After operation, we corrected the foils. During each operation, the positional relationship of each sample in the accelerator room was adjusted. The ^{198}Au activity of one gold foil was measured using a Ge detector (CANBERRA GC2518). The detection efficiency of gold foil measurements was calculated using ISOCS.[2,3] After measurement of the activity of one gold foil as a comparative standard, all foils were pasted on a sheet and contacted with an imaging plate (FUJIFILM Corporation, MS2025 or SR2025) to measure their relative activity simultaneously. We used Typhoon FLA7000 (GE Healthcare) to read the photostimulated fluorescence from imaging plates. Measurements were performed at two facilities with the same HM-18 accelerator (Sumitomo Heavy Industries, Ltd.): National Cerebral and Cardiovascular Center (NCCC), Japan, and Osaka University Hospital, Japan. At NCCC, H^- was accelerated (18 MeV) to produce ^{18}F and ^{11}C. At Osaka University Hospital, H^- (18 MeV) was used to produce ^{18}F, and ^{11}C and D^- (9 MeV) were used to produce ^{15}O. Figure 1 shows the location of gold foils in operation at Osaka University Hospital. Table 1 summarizes the operational conditions.

Table 1 *Operational parameters.*

Nuclide	^{18}F	^{18}F	^{11}C	^{11}C	^{15}O
Facility	NCCC	Osaka Univ	NCCC	Osaka Univ	Osaka Univ
Beam	H-	H-	H-	H-	d-
Energy [MeV]	18	18	18	18	10
Current [μA]	20	25	20	25	18

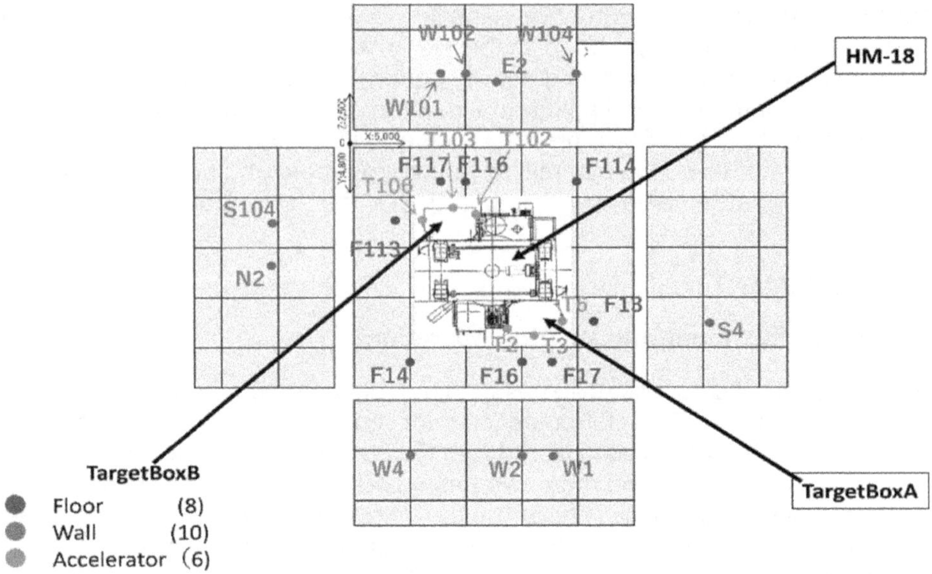

Figure 1 *Locations of gold foils at Osaka University Hospital. There are two target boxes. The target box used depends on the nuclide to be produced.*

2.2 Concrete core samples

At NCCC, we obtained concrete core samples from the locations at which gold foils were pasted. The diameter of each core was 65 mm and sampling depth was ~500 mm. From the surface to the bottom, 20 mm thick disks were cut at certain intervals to obtain measurement samples. Each disk was powdered and placed in a plastic vial (U-8). Samples were measured using a Ge-detector (ORTEC GMX-20195-S) to obtain the specific activity of each nuclide. Figure 2 shows the locations at which concrete core samples were obtained at NCCC.

3 RESULT AND DISCUSSION

3.1 Neutron flux monitoring during operation

We compared the neutron fluxes obtained at 11 locations with the same positional relationship to the target box in the accelerator room for each operation. The neutron flux during ^{18}F production was higher than that during ^{11}C and ^{15}O production. The maximum flux was measured on the cyclotron surface (Target Box B). The thermal neutron flux was 5.56×10^6 [s^{-1} cm^{-2}] and epithermal neutron flux was 1.82×10^5 [s^{-1} cm^{-2}] at point T102 in Figure 1. The beam current was different in each operation. Therefore, the beam current was normalized and the neutron flux in each operation mode was compared. The thermal neutron flux ratio for the production of ^{18}F, ^{11}C, and ^{15}O was 1:0.24:0.56 at F17 in Figure 1. Figure 3 shows the thermal neutron flux at this point, and Figure 4 shows the epithermal neutron flux at this point.

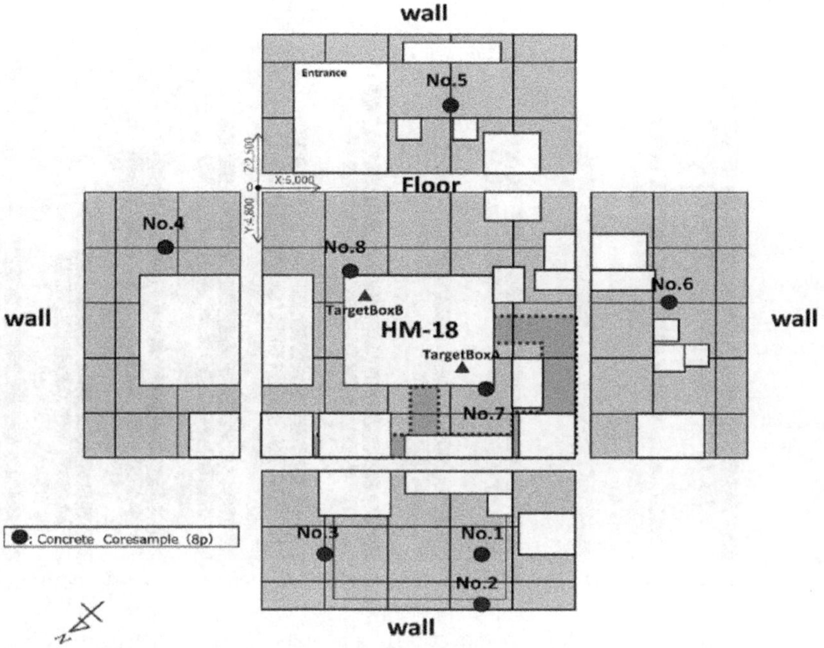

Figure 2 *Locations of concrete core samples at NCCC.*

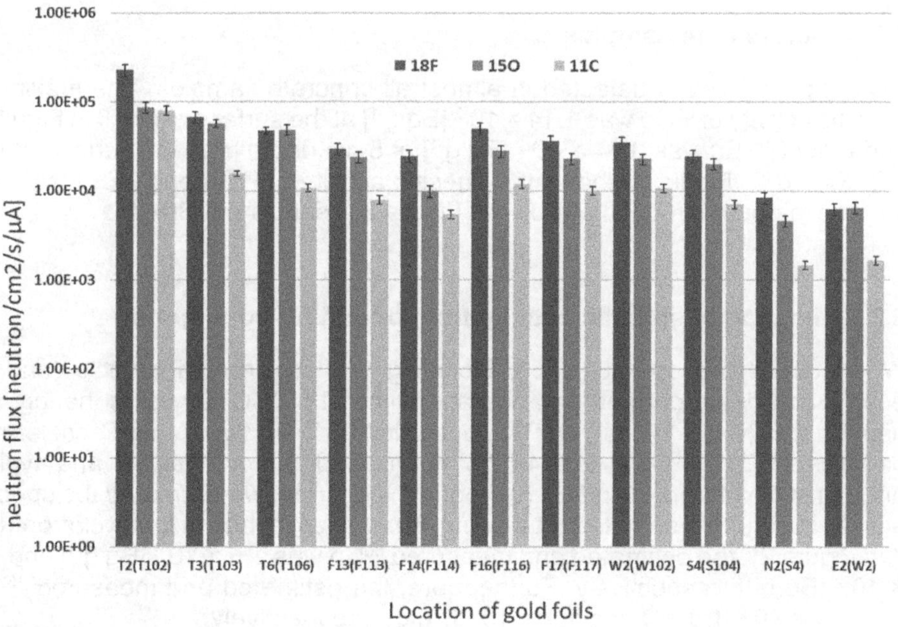

Figure 3 *Thermal neutron flux at the Osaka-University operations.*

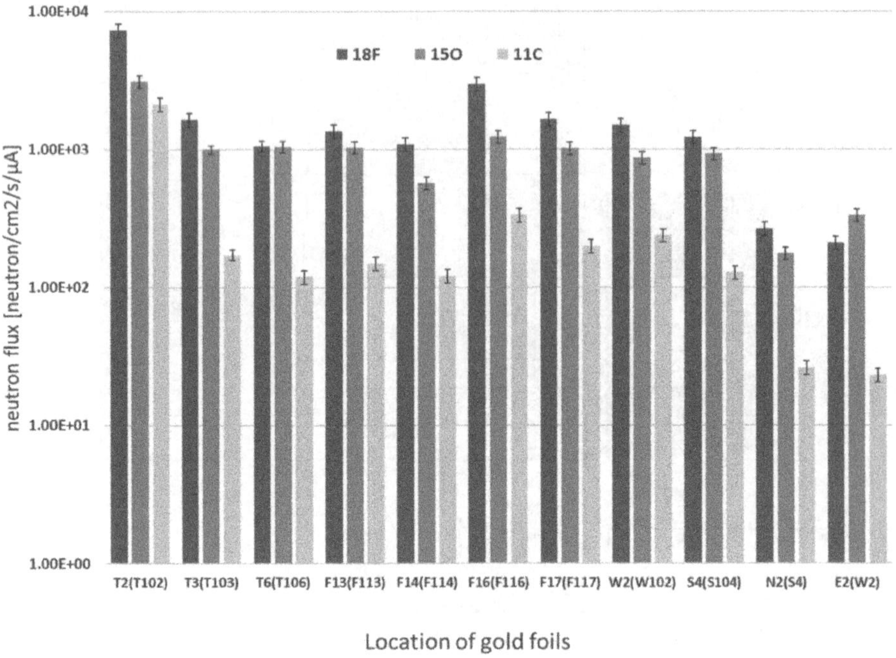

Figure 4 *Epithermal neutron flux at the Osaka-University operations.*

3.2 Concrete core samples

^{60}Co and ^{152}Eu were detected in almost all concrete samples. The maximum specific activity of ^{60}Co was 1.14 × 10^{-2} [Bq g^{-1}] at the surface of No. 6 in Figure2, and that of ^{152}Eu was 3.84 × 10^{-2} [Bq g^{-1}] at 5 cm depth from the surface of No. 7 in Figure 2. Figure 5 shows the specific activity of ^{60}Co and ^{152}Eu from the surface to depth direction at No. 7 in Figure 2 (near Target Box A).

3.3 Estimation of specific activity from operating conditions

We obtained the operation log, including the accelerated particle (and its energy and beam current) and operating time at NCCC. By using the log, the neutron flux (7.8 × 10^5 [s^{-1} cm^{-2}] in operation for ^{18}F), flux ratio of accelerated particles (^{18}F:^{11}C:^{15}O = 1:0.24:0.56 normalized current value), and typical amount of Co (10 ppm) and Eu (0.5 ppm) in concrete, we estimated the specific activity in concrete samples. At a point near the target box of the cyclotron (No. 7 in Figure 2), the estimated and measured ^{60}Co was 1.5 × 10^{-2} [Bq g^{-1}] and 1.1 × 10^{-2} [Bq g^{-1}], respectively. Furthermore, the estimated and measured ^{152}Eu was 3.1 × 10^{-2} [Bq g^{-1}] and 1.9 × 10^{-2} [Bq g^{-1}], respectively.

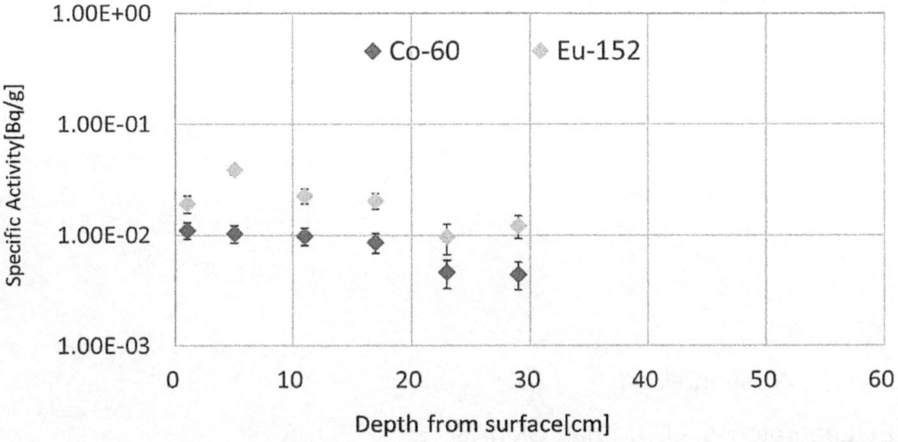

Figure 5 *Specific activity of* ^{60}*Co and* ^{152}*Eu from surface of concrete core No.7.*

4 CONCLUSIONS

The neutron flux attributable to accelerator operation at the PET cyclotron and residual radioactivity in the concrete structure could be determined nondestructively. The results of the nondestructive method agree with those of the sampling method. We found that radioactivity in the concrete structure can be evaluated easily if information of the flux ratio of operation mode and operation history is available.

CONFLICTS OF INTEREST

There are no conflicts to declare.

ACKNOWLEDGEMENTS

This research was performed with the aid of the Radiation Safety Research Promotion Fund “Clearance of materials from the decommissioning of accelerator facilities” by the Japan Nuclear Regulation Authority.

REFERENCES

1. H. Matsumura *et al.*, *IOP Conf. Series: Journal of Phys. Conf. Series*, 2018, **1046**, 012016.
2. H. W. Patterson and R. H. Thomas, *Accelerator Health Physics,* Academic Press, Inc., New York, 1973.
3. R. Venkataraman, F. Bronson, V. Atrashkevich, M. Field and B. M. Young, Improved detector response characterization method in ISOCS and LabSOCS, *J. Radioanal. Nucl. Chem.*, 2005, **264**, 213–219.

RADIUM MONITORING AT COAL FIRED POWER PLANTS

James B. Westmoreland

GEL Laboratories, LLC, Charleston, SC 29407, USA

jbw@gel.com

ABSTRACT

GEL Laboratories, LLC (GEL) supplies analytical testing services to a variety of industries. Recently GEL has observed an increase in the monitoring of naturally occurring radionuclides in the environment from non-nuclear customers including coal, steel, natural gas and the petroleum industries. Regulatory oversight for the coal industry in the United States has escalated the monitoring of radium and other elements around active and inactive coal ash storage basins. This paper shares specific results encountered in the monitoring process and considerations for radium method selection.

1 INTRODUCTION

Coal fired power plants have provided power around the world since the late 19th century. In 1882 the world's first coal fired power station was built by Edison Electric in London and coal-based power plants are still a major percentage (~40%) of global power production. The negative environmental aspects of coal-based power are also well documented. Solid waste is a major storage and disposal issue for utilities and a 2008 accident in Kingston

Tennessee USA released 4.2 million cubic meters of fly ash into a river basin causing massive damage. This incident accelerated US regulations for active and inactive coal plants known as the "coal combustion residuals rule" (CCR) (Ref.1).

The CCR addresses the storage of wastes and monitoring for chemical hazards that can aggregate in the coal cycle. The CCR Appendix IV (Table 1) specifies the elements for assessment in waste streams and monitoring wells around coal plants. The regulation assigns a combined maximum concentration for Radium 226 (^{226}Ra) and Radium-228 (^{228}Ra) as less than 0.18 Bq/L (5 pCi/L) in monitoring wells.

Table 1 *CCR Appendix IV.*

Antimony	Lithium
Arsenic	Mercury
Barium	Molybdenum
Beryllium	Selenium
Cadmium	Thallium
Chromium	Fluoride
Cobalt	**Radium 226 & 228**
Lead	pH (field)

Testing for these specified elements was undertaken by GEL in ground water monitoring wells, and solid waste fly ash. Solid samples were dissolved using a total mineral acid dissolution including nitric and hydrofluoric acids. ^{226}Ra was determined by a classical radon purge and trap method utilizing Lucas cell counters (Figure 1). ^{228}Ra determination followed a traditional chemical separation followed by DGA resin (Ref. 2) to isolate and count Actinium-228 (^{228}Ac) by gas flow proportional counting.

Several alternate methods are valid for these measurements and selected laboratory methods should take into consideration the potential for elevated radon, stable calcium and barium in ground water.

Alpha spectroscopy is an effective alternative method to radon emanation for a more rapid determination of ^{226}Ra. One challenge with the alpha spectroscopy method is the potential for poor peak resolution due to excess mass precipitated and filtered for counting as barium sulphate. In addition to resolution issues, ^{226}Ra daughters can also cause detector contamination and interferences.

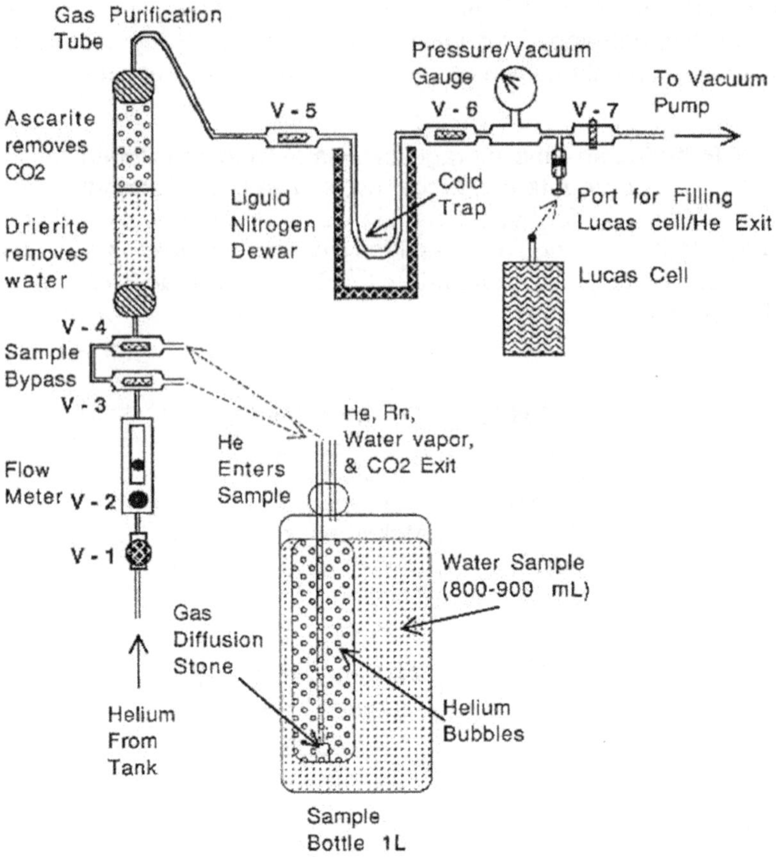

Figure 1 *De-emanation "Purge and Trap" Methodology for the Determination of ^{226}Ra.*

For solid samples, gamma spectroscopy is a good alternative with a simultaneous measurement of ^{226}Ra and ^{228}Ra along with other naturally occurring isotopes (Potassium-40 (^{40}K), Uranium, Thorium, etc.). The gamma method for solids allows for more actual sample volume to be tested (~120 grams) as compared to the classical methods for total dissolution using ~1g of material. The gamma method is most effective when samples can be sealed and allowed to equilibrate with Bismuth-214 (^{214}Bi), Lead-214 (^{214}Pb) and ^{228}Ac, the higher yield gamma daughters of ^{226}Ra and ^{228}Ra. In practice GEL has found unacceptably high uncertainty with attempting to measure the 186 keV ^{226}Ra gamma emission directly by gamma spectroscopy.

2 EXPERIMENTAL

The ^{226}Ra method was verified using three separate uranium mill tailings reference materials prepared by the Grand Junction Projects Office [Ref.3 & 4]. An average recovery of 97% was observed for ^{226}Ra as compared to the known values in Table 2.

Table 2 *^{226}Ra Solid Matrix Method Verification Data.*

SAMPLE	Result (Bq/kg)	CSU Ref. 5	MDC Ref.6	^{226}Ra Reference Value	% Recovery
NRM-2	37	16	8	41	90%
NRM-2 (dup)	38	16	13	41	92%
NRM-6	1703	399	15	1650	103%
NRM-6 (dup)	1604	378	14	1650	97%
TRM-3	2132	502	13	2231	96%
TRM-3 (dup)	2291	545	8	2231	103%

The ^{228}Ra method was verified by comparing NRM-6 to gamma spectroscopy results. While no reference value was available for ^{228}Ra, GEL gamma analysis revealed a significant ^{228}Ra present for comparison.

Table 3 *^{228}Ra Solid Matrix Method Verification Data.*

SAMPLE	Result (Bq/kg)	CSU	MDC	^{228}Ra Gamma Value	% Recovery
NRM-6	1207	188	38	1260	96%

Solid samples were analysed using a sample mass of 0.5 grams and count times of 30 minutes for Lucas cell counting and 60 minutes for gas flow proportional counting. Method volumes and count times resulted in an average sample specific MDC of 12.8 Bq/kg for ^{226}Ra and 28.6 Bq/kg for ^{228}Ra. Liquid samples were analysed using 500 mL for ^{226}Ra and 900 mL for ^{228}Ra. Counting intervals for ^{226}Ra were 30 minutes while ^{228}Ra was 60 minutes. These conditions produced an average MDC of 0.02 Bq/L for both ^{226}Ra and ^{228}Ra.

3 RESULTS

Samples were received weekly throughout the time interval from January to July. Water samples varied from 40-120 samples in a typical week. Solid samples were received on a much lower frequency of less than 10 samples per week. Batches of 20 or less samples were created and assigned to the laboratory analyst with an associated blank, laboratory control sample and duplicate. Confirmation analysis was performed when water sample results exceeded the 0.18 Bq/L CCR limit.

Table 4 *Summary of Study Results.*

Isotope	Matrix	Total Samples	Average Result	Std. Dev.	Maximum Result
^{226}Ra	Water	1100	0.05 Bq/L	0.09	1.8 Bq/L
^{228}Ra	Water	1100	0.02 Bq/L	0.05	0.5 Bq/L
^{226}Ra	Solid	80	30 Bq/kg	16	90 Bq/kg
^{228}Ra	Solid	80	22 Bq/kg	18	90 Bq/kg

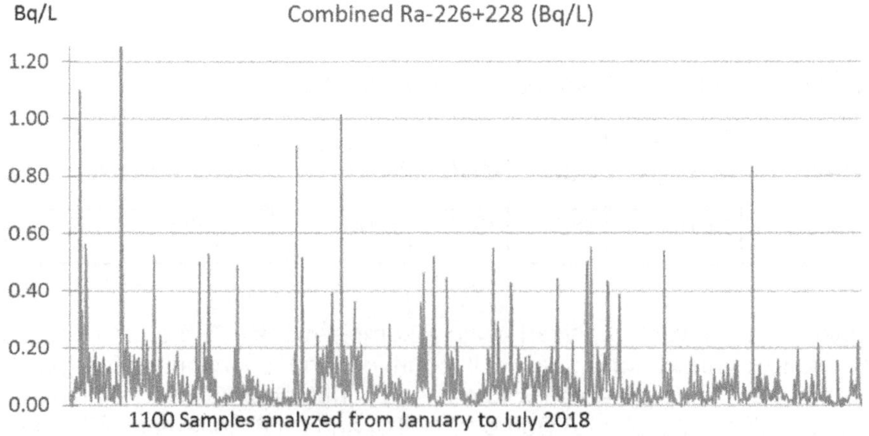

Figure 2 *Graphical Summary of Combined Radium Study Results.*

4 DISCUSSION

Approximately 7% of 1100 ground water samples tested between January and July 2018 exceeded the CCR combined radium limit. No consistent ratios were observed between ^{226}Ra and ^{228}Ra when elevated results were observed, however, ^{226}Ra was more frequently the cause of exceedances in the total data set. The primary interference encountered for ^{226}Ra by the de-emanation method was elevated natural radon in monitoring wells. This interference was mitigated by retaining samples after the initial de-emanation cycle for reanalysis 3-5 days later. Rapid declines in activity were an indicator of radon interference while consistent results confirmed the presence of ^{226}Ra.

The question of reporting uncertainties (CSU) and detection limits (MDC) was raised since the CCR regulation combines the results of two separate methods. Simply adding the results together is acceptable but adding the CSU or MDC values is not a good practice. GEL opted for adding the two results and summing the squares of the two CSU values. GEL recommended reporting both MDC values or using the higher value between the methods as a conservative solution.

Solid samples were sampled on a lower frequency and primarily consisted of fly ash. Interest in the fly ash was for modelling ash ponds and comparing results with monitoring wells. The total dissolution method achieved reasonable recoveries above 96% for both ^{226}Ra and ^{228}Ra as compared with reference materials noted in Ref. 3 and 4.

5 CONCLUSIONS

For this project using radon emanation for ^{226}Ra and gas flow proportional counting for ^{228}Ra was effective. Due to the high continued demand for this testing, subsequent work has shifted to analysing ^{226}Ra as total alpha radium by gas flow proportional method based on EPA Method 903.0. This screening method is a simple bulk barium sulphate precipitation technique where the final counting is done in the gross alpha window of a gas flow proportional counter. This method allows for increased sample throughput and the de-emanation method is used to confirm ^{226}Ra if positive results are observed for the total alpha radium technique.

Method selection is vital for successful monitoring of radium. Balancing technical and practical considerations will help ensure quality and capacity considerations are met.

ACKNOWLEDGEMENTS

Radon emanation schematic provided with thanks to Dr William Burnett for this design from Florida State University Department of Oceanography.

NOTES AND REFERENCES

1. Coal Combustion Residuals Rule, US Environmental Protection Agency, 80 Federal Register 21301, Agency Docket Number EPA–HQ–RCRA–2009–0640, 2015.
2. E. P. Horwitz, D. R. McAlister, A. H. Bond, R. E. Barrans, Novel Extraction Chromatographic Resins Based on Tetraalkyldiglycolamides: Characterization and Potential Applications, Solvent Extraction Ion Exch., 23, 219 (2005). **(HP104)**
3. S. Donivan and R. Chessmore, Uranium Reference Materials, DOE/ID/12584-9, 1987.
4. S. Donivan and R. Chessmore, Soil Based Uranium Disequilibrium and Mixed Uranium-Thorium Radionuclide Reference Materials, DOE/ID/12584-34, 1988.
5. Multi-Agency Radiological Laboratory Analytical Protocols Manual (MARLAP) (NUREG-1576), Chapter 19 (2004). CSU is an acronym for Combined Standard Uncertainty. It incorporates a one sigma counting uncertainty along with other sources of analytical method uncertainties.
6. Multi-Agency Radiological Laboratory Analytical Protocols Manual (MARLAP) (NUREG-1576), Chapter 20, 2004. MDC is an acronym for Minimum Detectable Concentration. This calculation represents a sample specific detection limit utilizing a 95% confidence interval.

EVALUATION OF DIFFERENT GAMMA-RAY IMAGING TECHNIQUES FOR VISUALISATION OF INDUCED ACTIVITY IN ACCELERATOR MAGNETS

Go Yoshida*, Akihiro Toyoda, Hiroshi Matsumura, Kazuyoshi Masumoto, Taichi Miura, Hajime Nakamura, and Kotaro Bessho

High Energy Accelerator Research Organization (KEK) 1-1 Oho Tsukuba Ibaraki Japan

yoshigo@post.kek.jp

ABSTRACT

Activated accelerator magnets were measured with three different types of commercially available gamma-ray imaging devices (pinhole, Compton scattering, and masked type) and the results were compared to simplify a distinction between activated/non-activated areas in the magnet during the decommissioning of an accelerator facility. In general, an activated magnet includes ^{60}Co and emits gamma rays over 1 MeV; however, current devices are not designed to detect such high-energy photons. Our results showed that two devices could identify the location of highly activated areas, although they could not detect the total absorption peaks of the 1173 and 1333 keV photons from ^{60}Co.

1 INTRODUCTION

Accelerator magnets are activated by beam loss during operation.[1–3] When an accelerator facility is being decommissioned, it is necessary to evaluate the activated area of the magnet to determine if reuse is possible and if radiation protection for workers is required. If the probe of a survey meter is moved away from the radiation source, the indication value of the dose rate decreases radically. In addition, an activated magnet cannot be regarded as a point source, but rather as a source with contrast and distribution. Under these conditions, unless a survey meter with an appropriate shield is used, we cannot identify the activated areas and strength precisely. Therefore, it takes significant effort and time to distinguish activated / non-activated areas of an accelerator magnet.

Since the accident at the Fukushima Daiichi Nuclear Power Plant, the gamma-ray imaging technique has rapidly advanced. This technique visualises relatively high surface dose areas and rapidly provides a wide range of information on the activation. However, the evaluation of activation in accelerator facilities by this gamma-ray imaging technique is rarely performed. In cases of activation in accelerator facilities, ^{60}Co, which emits high-energy (1173 and 1333 keV) photons, is the principal nuclide, but most current techniques aim for the detection of ^{137}Cs, which emits 662 keV photons. Additionally, because entire sections of the facility such as the walls, floor, and ceiling become activated and emit gamma rays,[4–6] identifying the source location is quite difficult.

In this study, we evaluated and compared the current imaging techniques based on the following criteria: 1) whether detection of gamma rays of ^{60}Co is possible, and 2) whether accurate detection of the location of the radiation source is possible. To evaluate the induced activation of the accelerator magnet with the gamma-ray imaging technique, measurements of activated magnets employing commercially available devices were performed in coordination with the manufacturer engineers, and the results were compared.

2 EXPERIMENTAL

2.1 Activated magnets

Four activated magnets, which were installed in the 12 GeV proton beam line in the KEK Tsukuba, and a magnet, which was installed in the 6 GeV electron beam line, were adopted as objects. To reduce the adverse effects of strong radiation from the beam lines on the measurements, these magnets were moved from the accelerator tunnel to a storage facility for activated materials. An ambient dose rate measurement was conducted before the installation of all magnets and it was confirmed that there were no high-dose materials around magnets. The dose rate of the measurement site was 0.1–0.5 μSv/h. Information for each of the test magnets, including the maximum surface dose rate, detected nuclides, and installation location, is summarised in Table 1. The surface of the magnets and the water-cooled piping were flushed with water, and we confirmed the surface contamination with smear test.

Before measurement with imaging devices, we measured the magnets with an ionisation-chamber survey meter, ICS-321 (Hitachi Aloka Medical Ltd., Japan), and a Zinc-doped Cadmium Telluride (CZT) spectrometer, GR-1 (Kromek, UK), for surface dose rate and nuclide identification, respectively. The principal nuclides in these magnets were ^{60}Co, ^{22}Na, and ^{54}Mn. Details are provided in the supplementary information.

Table 1 *Activated accelerator magnets that were adopted as the objects of gamma-ray imaging measurement.*

Name	Magnet-A	Magnet-B	Magnet-C	Magnet-D	Magnet-E
Appearance					
Maximum surface dose rate	110 μSv/h	15 μSv/h	40 μSv/h	20 μSv/h	1 μSv/h
Detected nuclides	^{60}Co, ^{22}Na	^{60}Co, ^{22}Na	^{60}Co, ^{22}Na	^{60}Co, ^{22}Na	^{60}Co, ^{54}Mn
Installation	12 GeV, p^+ line	12 GeV, p^+ line	12 GeV, p^+ line	12 GeV, p^+ line	6 GeV, e^- line
Operation period	1990~2005	1999~2004	1999~2004	1999~2005	~2016

2.2 Gamma-ray imaging devices

Three different types of gamma-ray imaging devices, namely the HGD-E1500 (Hitachi Ltd., Japan),[7] the Gamma-Catcher (Chiyoda Technol Corporation, Japan),[8] and the iPIX (Mirion Technologies Canberra, US)[9] were employed. These were selected as representatives of pinhole, Compton scattering, and masked types, respectively, which are the currently available imaging technologies. The HGD-E1500 uses Cadmium Telluride (CdTe) crystals as a gamma-ray detector. Lead shielding that is 4 cm thick surrounds the detector section, rendering it less susceptible to background radiation than other detectors. The Gamma-Catcher uses Gadolinium Aluminium Gallium Garnet (GAGG) crystals as a detector.

This device is lightweight and has better angular resolution than HGD-E1500. This device can reflect an arbitrary energy region of photons to the visualisation image. The iPIX is the smallest and lightest of the three models. It uses CZT crystals as a gamma-ray detector. An imaging device expresses the distribution of the surface dose rate on the object as a relative value. Three devices can adjust the aperture value, that is, the relative display lower limit dose. By changing the aperture value, the contrast of the contour diagram changes. The HGD-E1500 and the Gamma-Catcher have a spectrometry mode, and an energy spectrum can be obtained simultaneously with the visualisation. Table 2 summarises the specifications for the devices used in this study.

Table 2 *Specification of gamma-ray imaging devices that were employed for the measurement of activated magnets in this study.*

Detection method	Pinhole	Compton Scattering	Masked
Product Name	HGD-E1500 [7]	Gamma-Catcher [8]	iPIX [9]
Manufacturer	Hitachi	Chiyoda Technol	Mirion Technologies Canberra
External dimensions	193×327×295 mm	138×150×150 mm	188×90×90 mm
Weight	32 kg	2 kg	2 kg
Detector crystal (Thickness)	CdTe	GAGG (5+10 mm)†	CZT (1 mm)
Shield	Equivalent to 40 mm of Pb	None	None

†Scattering-region and absorption-region thickness

3 RESULTS AND DISCUSSION

3.1 Comparison of visualisation images of gamma rays

Table 3 summarises the measurement results. A constant aperture value was applied for all conditions in order to make the image easy to compare. The measurement time and distance between an object and the device are included in each section. Gamma-ray spectrums for each HGD-E1500 (it can only display up to 1000 keV) and Gamma-Catcher measurement are shown next to their respective images.

Table 3 *Comparison of visualisation images of gamma rays emitted from the activated magnets. A gamma-ray spectrum is shown next to each visualised image for the HGD-E1500 and Gamma-Catcher.*

Object	HGD-E1500 (Pinhole)	Gamma-Catcher (Compton Scattering)	iPIX (Masked)
^{60}Co Standard source	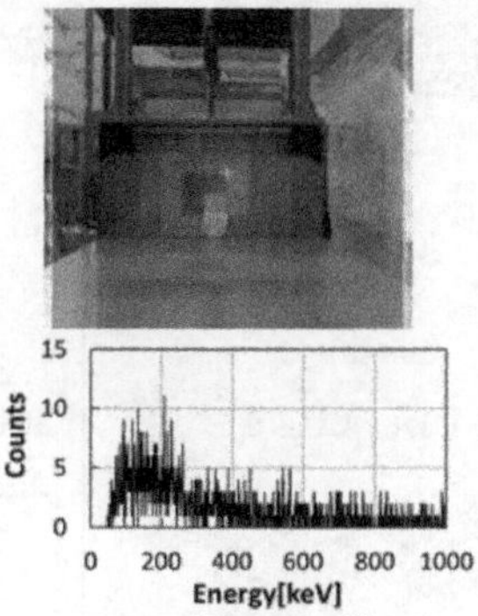 1 m*, 950 s**	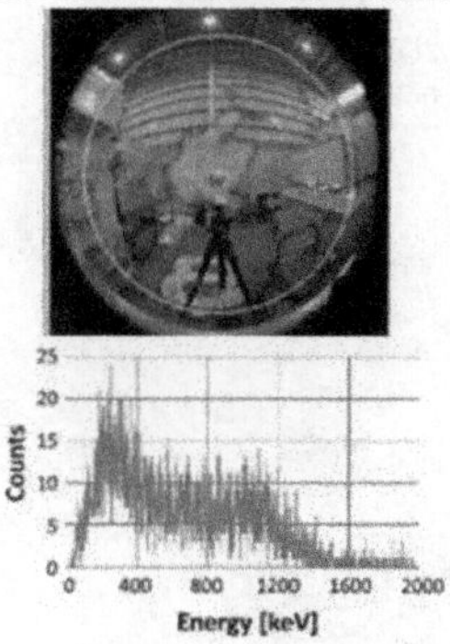 0.5 m*, 1055 s**	 0.5 m*, 594 s**
Magnet-A	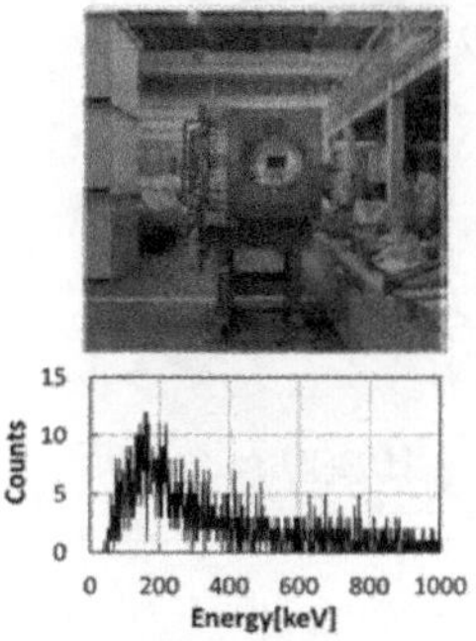 3 m*, 351 s**	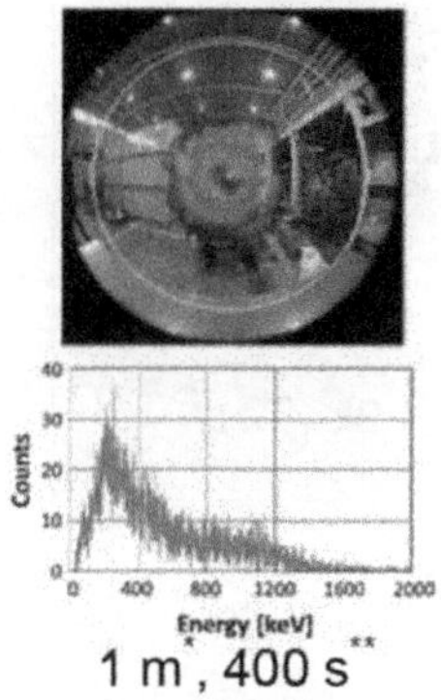 1 m*, 400 s**	 2 m*, 931 s**
	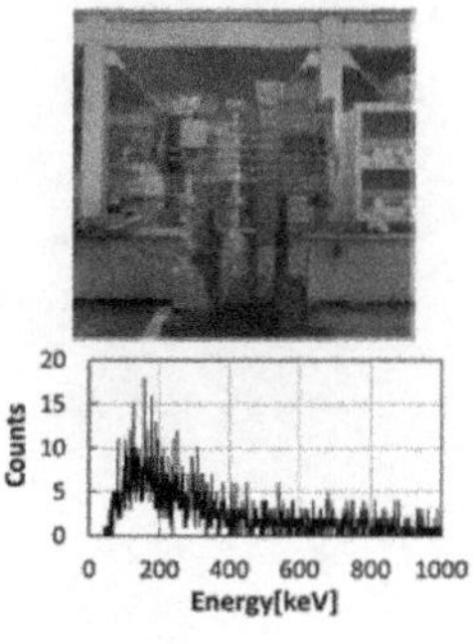 2.5 m*, 376 s**	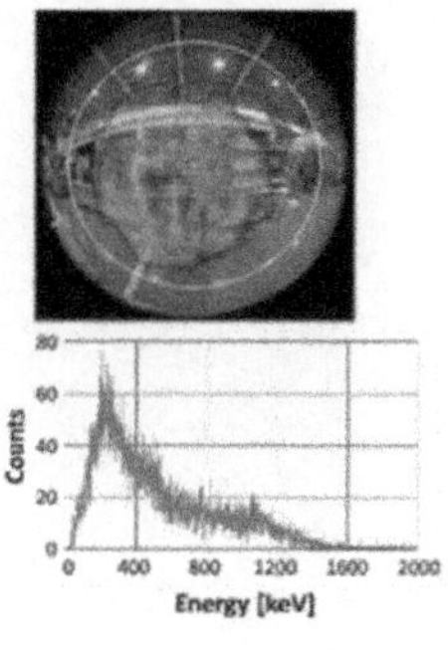 1 m*, 705 s**	 1.5 m*, 2151 s**

Object	HGD-E1500 (Pinhole)	Gamma-Catcher (Compton Scattering)	iPIX (Masked)
Magnet-B	 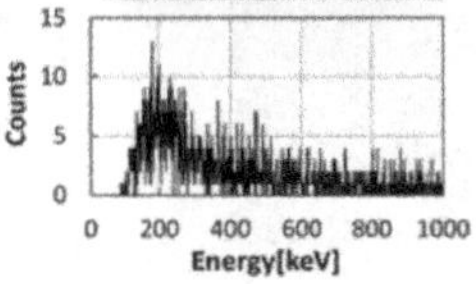 3 m[*], 1442 s[**]  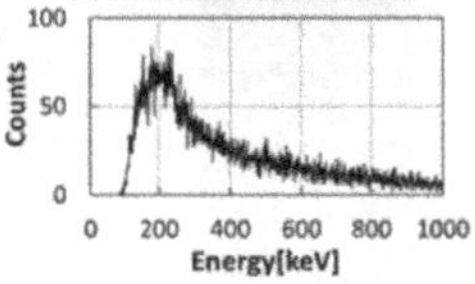 3 m[*], 21472 s[**]	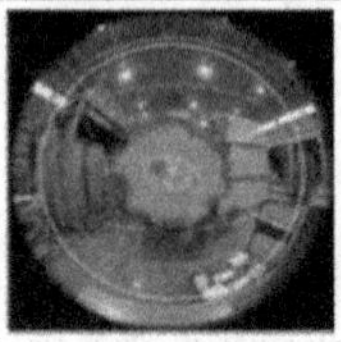 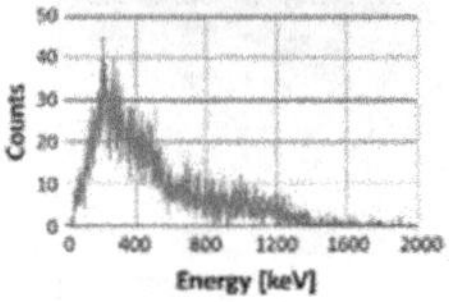 1 m[*], 602 s[**] 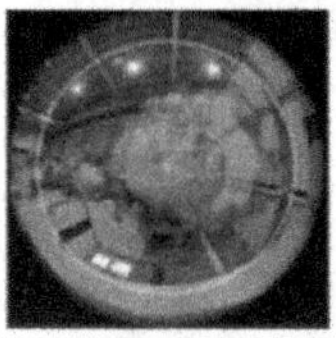 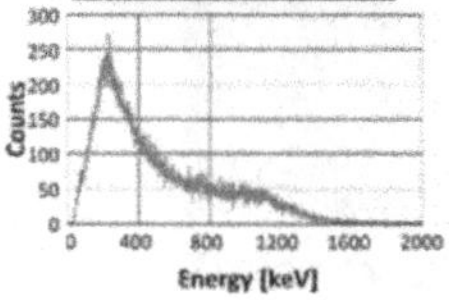 1 m[*], 10800 s[**]	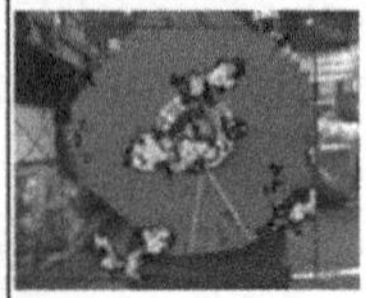 1.2 m[*], 3004 s[**] Not measured

<table>
<tr><th>Object</th><th>HGD-E1500 (Pinhole)</th><th>Gamma-Catcher (Compton Scattering)</th><th>iPIX (Masked)</th></tr>
<tr><td rowspan="2">Magnet-C</td><td>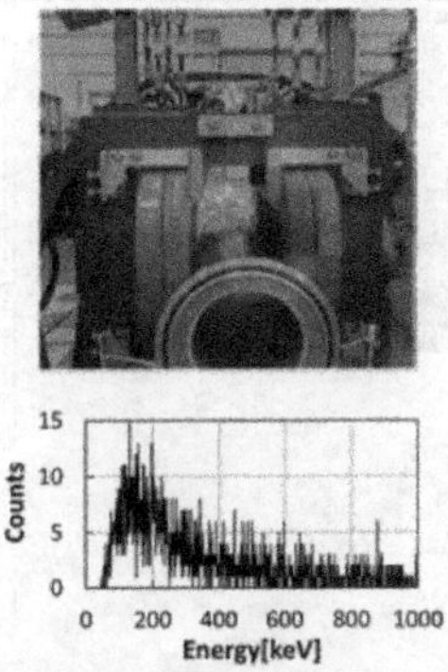

1 m*, 243 s**</td><td>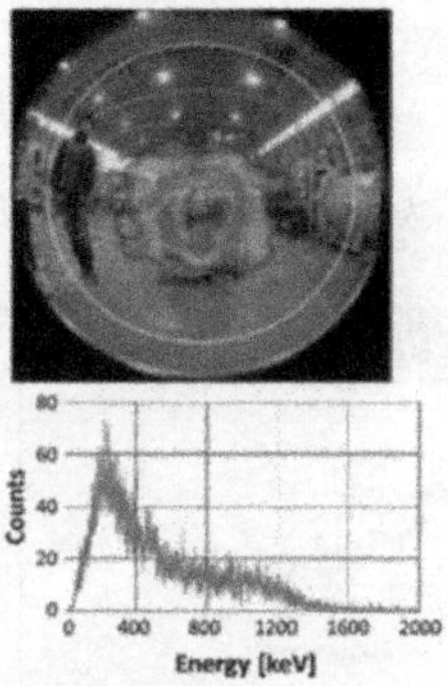

1 m*, 601 s**</td><td>
1 m*, 4271 s**</td></tr>
<tr><td>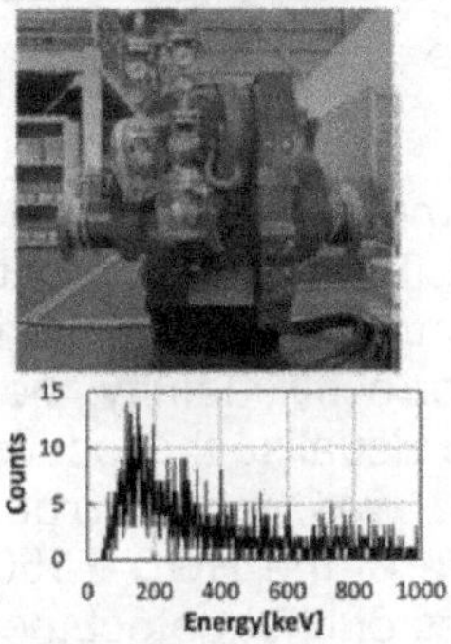

1.5 m*, 548 s**</td><td>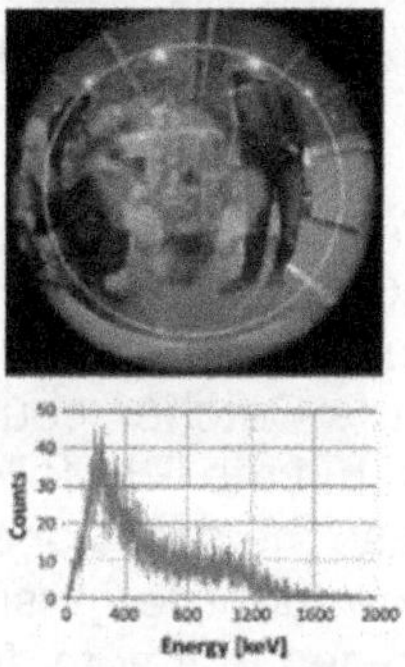

1 m*, 904 s**</td><td>Not measured</td></tr>
<tr><td>Magnet-D</td><td>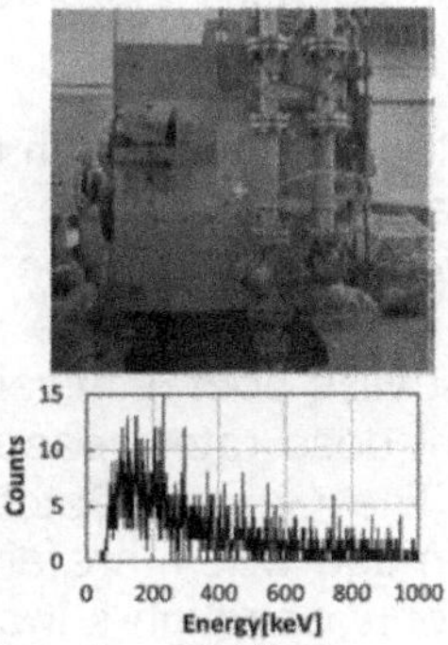

2 m*, 456 s**</td><td>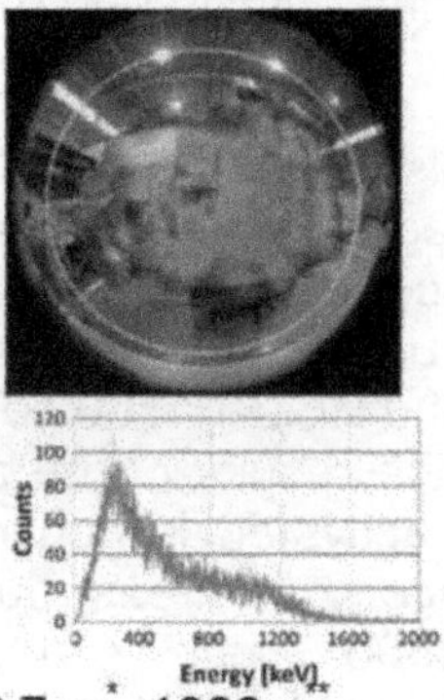

0.7 m*, 1000 s**</td><td>Not measured</td></tr>
</table>

Object	**HGD-E1500 (Pinhole)**	**Gamma-Catcher (Compton Scattering)**	**iPIX (Masked)**
Magnet-E	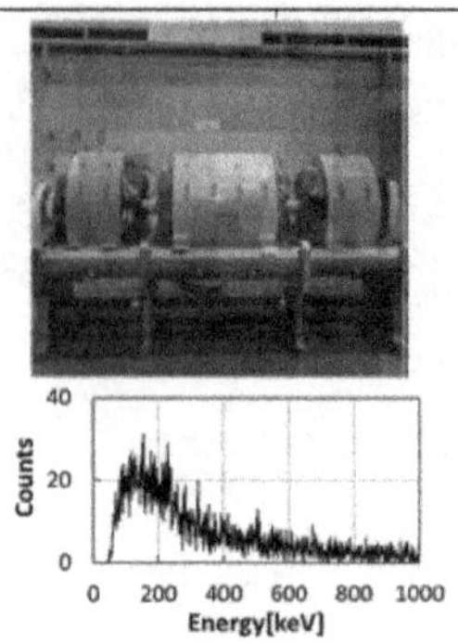 1 m*, 5872 s**	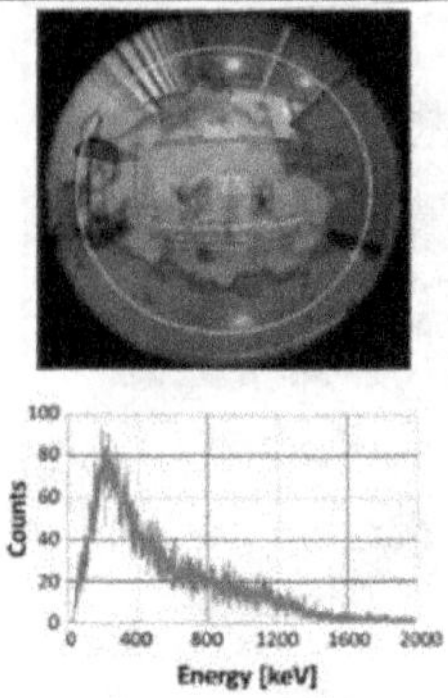 0.5 m*, 7200 s**	Not measured

*Distance between an object and the device
**Measurement time

3.1.1 Standard radiation source. First, we measured a ^{60}Co standard radiation source (500 kBq) to confirm the ability of the devices to detect high-energy gamma rays. The source was placed on the centre of the image. The HGD-E1500 and the Gamma-Catcher could correctly identify the location of the source and visualise it, but the iPIX failed to visualise the location of the source. This failure was because the detector crystal was too thin to detect high-energy gamma rays. Based on the energy spectra of the HGD-E1500 and Gamma-Catcher, it was determined that both devices did not detect the total absorption peaks of ^{60}Co; however, they were able to detect scattered photons with low energies. The estimated location of the source determined by the HGD-E1500 was located above the actual source location because the parallax could not be corrected.

3.1.2 Magnet-A. This is the strongest activated magnet. When directed at the front side of the magnet (upper image in Table 3), all devices visualised the relatively high surface dose area in the centre of the beam pipe, but the image from the iPIX was unclear. When directed at the side of the magnet (lower image in Table 3), the HGD-E1500 could visualise both relatively high surface dose areas. From the results of the ambient dose rate measurement, same regions are highly activated is confirmed. The Gamma-Catcher could also visualise high surface dose areas, but the image was more ambiguous than that of the HGD-E1500. The iPIX detected only one high surface dose area. For the HGD-E1500 and the Gamma-Catcher, images converged sufficiently in 6 to 12 minutes, whereas the iPIX did not obtain images equivalent to other models even with more time. Furthermore, convergence of the image with the iPIX was reduced when the distance to the subject was shortened to 1 m. In the energy spectra of the HGD-E1500 and the Gamma-Catcher, total absorption peaks were not confirmed. Similar to the measurement of the ^{60}Co

radiation source, it was determined that visualisation images were attributed to scattered photons from the magnet.

3.1.3 Magnet-B. When directed at the front side of the magnet (upper image in Table 3), the HGD-E1500 and the Gamma-Catcher visualised the relatively high surface dose area clearly on the centre of the beam pipe, but the iPIX could not correctly identify the high surface dose area because of insufficient detection efficiency. Confirmation that the contrast of the surface dose rate distribution for Magnet-B is not sharper than that for Magnet-A was obtained with preliminary ambient dose measurement. In an energy spectrum of the HGD-E1500, a slight prominence considered to be derived from ^{22}Na is confirmed around 511 keV, whereas no peaks could be confirmed with the Gamma-Catcher. The beam pipe of Magnet-B is made of aluminium alloy, and a significant amount of ^{22}Na is generated by proton or fast neutron reactions. The beam pipe of Magnet-A, however, is made of stainless steel. Therefore, it is harder to generate ^{22}Na than it is in the case of an aluminium beam pipe. Because there are also more scattering photons derived from ^{60}Co in Magnet-A, the 511 keV peak becomes relatively small. When directed at the side of the magnet (lower image in Table 3), the HGD-E1500 visualised high surface dose areas on the coolant pipes. The Gamma-Catcher also visualised high surface dose areas, but the image was less clear than that of the HGD-E1500. In this measurement, if the measurement time increased, we confirm that the resolution of the images did not improve. No peaks were identified in the energy spectrum by the HGD-E1500 or the Gamma-Catcher. A measurement with the iPIX was not performed owing to shortage of time. In addition, Magnet-C (side-view), Magnet-D, and Magnet-E were also not measured with the iPIX.

3.1.4 Magnet-C. This magnet was activated asymmetrically. Coils were the most strongly activated. In this measurement, the HGD-E1500 was placed 1 m from the object to improve resolution, even though the proper focal length is 5 m. When directed at the front side of the magnet (upper image in Table 3), the HGD-E1500 and the Gamma-Catcher visualised the high surface dose area on the coils. When directed at the side of the magnet (lower image in Figure 3), the HGD-E1500 visualised two high surface dose areas on the coil. However, the Gamma-Catcher could not separate the high surface dose areas. No peaks were confirmed in the spectrum by either device.

3.1.5 Magnet-D. This magnet was located downstream of a kicker magnet which changed the direction of the proton beam. A branch beam line had been installed at the side of this magnet. The kicker-installed side (left of the image) had a relatively high dose rate owing to the activation originating from the beam loss in the kicker magnet. Coolant piping was also relatively activated. When directed at the side of the magnet, both the HGD-E1500 and the Gamma-Catcher visualised the high surface dose area on the yoke, but neither highlighted the high dose area on the coolant piping. No peaks were confirmed in the spectrum by either device.

3.1.6 Magnet-E. This is the weakest activated magnet. The maximum surface dose rate of this magnet is 100 times lower than Magnet-A, and the contrast of the surface dose rate is not sharp on the whole. However, both the HGD-E1500 and the Gamma-Catcher could visualise high surface dose areas. Although the image of Gamma-Catcher shows the high dose area distributes as a whole, it is caused by the set value of aperture. In this measurement, it took more than 1 hour for the image to converge. No peaks were confirmed in the spectrum by either device.

3.2 Detection efficiency of devices

Based on measurement results, detection efficiency plays the most important role in the visualisation of high surface dose areas of activated accelerator magnets. The iPIX, which has the thinnest crystal, was mostly unable to visualise the high surface dose areas on the magnets, despite a longer measurement time. Conversely, detection efficiencies of the HGD-E1500 and the Gamma-Catcher were not sufficient for nuclide identification. In almost all conditions, peaks did not appear on the spectrum. Although detection efficiency is not disclosed for any of the devices, we estimated the efficiencies of the detectors with EGS5 code[10] and summarised the results in Table 4. We calculated with the condition of measuring the1333 keV point source from 1 m away for each device. For reference, the efficiency of the Lanthanum Bromide ($LaBr_3$, with crystal size of1.5 in. diameter × 1.5 in. length) scintillation-type spectrometer: InSpector 1000 (Mirion Technologies Canberra, US) was also calculated. As shown in Table 4, the efficiency of the Gamma-Catcher is less than 1/10 of $LaBr_3$; moreover, the efficiency of the iPIX is less than 1/100 of the $LaBr_3$. It was suggested that the threshold level to visualise a hot spot of ^{60}Co on the object is 1/10 of the efficiency of the $LaBr_3$ spectrometer. The InSpector 1000 $LaBr_3$ spectrometer is convenient to carry and we often use it for identification of nuclides in the induced activities of various accelerator components. The detection efficiency of the InSpector 1000 can be obtained with ISOCS,[11,12] and we confirmed that the calculation value of ISOCS and that of the EGS5 were consistent. No information about the detector crystal is disclosed for the HGD-E1500, thus the detection efficiency could not be calculated.

Table 4 *Relative detection efficiency of devices to InSpector 1000 $LaBr_3$ spectrometer, estimated with EGS5 code.*[10]

Device	**Relative detection efficiency**
Gamma-Catcher	0.055
iPIX	0.007
$LaBr_3$ spectrometer (ref.)	1.000

3.3 Effects of background gamma rays from the surroundings

To confirm the adverse effect of background gamma rays from surrounding materials on visualisation, simulated experiments employing two activated magnets were performed.

The HGD-E1500 was solely employed in this experiment, since extra time for short evaluation was left. Detailed information and results are summarised in Table 5. First, we placed two magnets (Magnet-A and B) next to each other and placed the imaging device in front of Magnet-B. In the visualisation image, a relatively high surface dose area shifted to the right compared to the reference (same as the image in Table 3), owing to gamma rays from Magnet-A. Next, we placed the device between Magnet-B and Magnet-C and measured Magnet-B. In this case, the visualisation image was disordered, and the relatively high surface dose area appeared at a different position than expected. Despite the radiation shield equivalent to 40 mm of lead on the HGD-E1500, the effects of background gamma rays were not completely suppressed. A thicker radiation shield would be required if it were to be used in a higher dose environment such as the beam line tunnel.

4 CONCLUSIONS

To investigate the activation of an accelerator magnet rapidly and comprehensively, a fundamental evaluation for applying a radiation imaging technique was performed. We measured activated magnets with three different gamma-ray imaging devices (pinhole, Compton scattering, and masked) as representatives of commercially available devices. The results of this study show that the detector crystal thickness and density, i.e. the detection efficiency, is more important than the detection method. Additionally, detection of the total absorption peak of ^{60}Co is not always necessary to distinguish and visualise the high surface dose area of an activated magnet. The detection of scattered photons from the activated magnet can, to some extent, reproduce the actual conditions. Currently, we are planning to perform measurements of an accelerator magnet in a beam line tunnel to investigate the influence of radiation from the surroundings in detail. We will clarify the issues of the gamma-ray imaging of activated accelerator magnets for practical on-site operation.

CONFLICTS OF INTEREST

There are no conflicts to declare.

ACKNOWLEDGEMENTS

This research was performed with the aid of the Radiation Safety Research Promotion Fund, "Clearance of materials from the decommissioning of accelerator facilities", by the Japan Nuclear Regulation Authority.

Table 5 *Results of simulated experiments to confirm the adverse effect of background gamma rays from surrounding materials.*

<table>
<tr><th>Condition</th><th>Result</th></tr>
<tr><td>The device is located in front of Magnet-B (ref.)
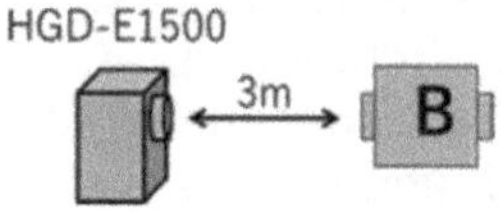
</td><td>
3 m*, 1442 s**</td></tr>
<tr><td>Magnet-A and B were next to each other
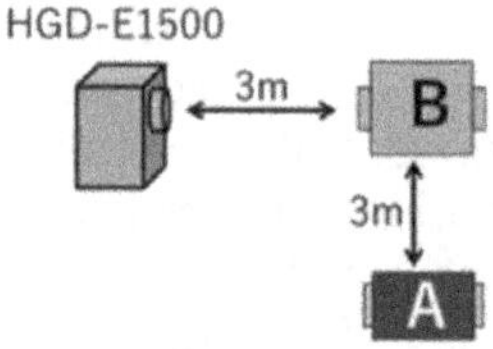
</td><td>
3 m*, 687 s**</td></tr>
<tr><td>Device was located between Magnet-B and C
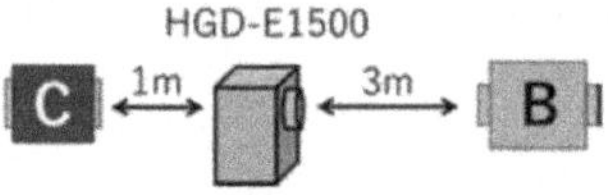
</td><td>
3 m*, 759 s**</td></tr>
</table>

*Distance between an object and the device
**Measurement time

NOTES AND REFERENCES

1. C. Bungau, A. Bungau, R. Cywinski, R. Barlow and T. R. Edgecock, *Phys. Rev. Spec. Top.–Accel. Beams*, 2014, **17**, 084701.
2. M. Brugger, A. Errahhaoui, M. Kim, H. S. Lee, S. Roesler and H. Vincke, *Progress in Nuclear Science and Technology*, 2014, **4**, 363.
3. A. Toyoda, G. Yoshida, H. Matsumura, K. Masumoto, T. Nakabayashi, T. Yagishita and H. Sasaki, *J. Phys. Conf. Ser.*, 2018, **1046**, 012017.
4. K. Masumoto, H. Matsumura, K. Bessho and A. Toyoda, *J. Radioanal. Nucl. Chem.*, 2008, **278**, 449.
5. H. Matsumura, A. Toyoda, K. Masumoto, G. Yoshida, T. Yagishita, T. Nakabayashi, H. Sasaki, K. Matsuura, Y. Yamaya and Y. Miyazaki, *J. Phys. Conf. Ser.*, 2018, **1046**, 012016.
6. I. L. Rakhno, N. V. Mokhov, I. S. Tropin and D. Ene, *J. Phys. Conf. Ser.*, 2018, **1046**, 012020.
7. Hitachi, Ltd. Online, http://www.hitachi.co.jp/products/healthcare/products-support/radiation/gammacamera/gammacamera/index.html, (accessed November 2018, Japanese only).
8. Chiyoda Technol Corporation Online, http://www.c-technol.co.jp/nuclear_power/power09, (accessed November 2018, Japanese only).
9. Data sheet of iPIX on the Mirion Technologies Online, https://mirion.s3.amazonaws.com/cms4_mirion/files/pdf/spec-sheets/ipix-ultra-portable-gamma-ray-imaging-system.pdf?1523467513, (accessed November 2018).
10. H. Hirayama, Y. Namito, A. F. Bielajew, S. J. Wilderman and W. R. Nelson, *The EGS5 Code System. SLAC-Report-730*, 2005.
11. R. Venkataraman, F. Bronson, V. Atrashkevich, M. Field and B. M. Young, *J. Radioanal. Nucl. Chem.*, 2005, **264**, 213.
12. F. L. Bronson, *J. Radioanal. Nucl. Chem.*, 2003, **255**, 137.

SUBJECT INDEX

ISOTOPES INDEX